Wireless Sensors and Instruments

Networks, Design, and Applications

Wireless Sensors and Instruments

Networks, Design, and Applications

Halit Eren

Taylor & Francis

Taylor & Francis Group

Boca Raton   London   New York

A CRC title, part of the Taylor & Francis imprint, a member of the
Taylor & Francis Group, the academic division of T&F Informa plc.

Published in 2006 by
CRC Press
Taylor & Francis Group
6000 Broken Sound Parkway NW, Suite 300
Boca Raton, FL 33487-2742

Library of Congress Cataloging-in-Publication Data

Eren, Halit, Ph. D.
    Wireless sensors and instruments : networks, design, and applications / Halit Eren.
        p. cm.
    Includes bibliographical references and index.
    ISBN 0-8493-3674-0 (alk. paper)
    1. Wireless communication systems--Design and construction. 2. Electronic instruments--Design
and construction. 3. Sensor networks--Design and construction. I. Title.

TK5103.2.E74 2005
621.384--dc22   2005050502

Taylor & Francis Group
is the Academic Division of Informa plc.

Visit the Taylor & Francis Web site at
http://www.taylorandfrancis.com

and the CRC Press Web site at
http://www.crcpress.com

Dedication

Dedicated to Semra, Ahmet, Suheyla, and Duygu

Preface

The evolutionary tree representing the growth of instruments and instrumentation technology is marked by a major fork. One branch is representing stand-alone instruments fulfilling the tasks of the requirements of dedicated measurements. Another branch is represented by instruments that can be networked to cooperatively work with many others to measure the variables of complex systems. Neither of these branches is new. What is new is that the networking of instruments can be done without the need for physical hard-wired connections. Wireless connectivity and networkability opens up many possibilities of research, development, and applications that we could not even dream of in the near past. Instruments can now be networked by new and novel techniques while they are on the move in their individual environments performing their tasks.

Many of us are very familiar with a number of wireless devices ranging from mobile and cordless telephones, pagers, garage door openers, remote controllers, home entertainment equipment controllers, and so on. The proliferation of wireless systems in consumer products and industrial applications is so fast that a progressive involvement in technology development is vital for many governments and industrial establishments to maintain competitiveness. This book will be helpful to highlight the expansion of wireless systems in instrumentation and measurements.

Instrumentation requires a broad knowledge involving a diverse range of disciplines, such as measurement science, design and construction of electronic circuits, understanding of IC technology, appreciation of theory and practice of wireless communication systems, networks, protocols, and standards. This book is specifically written to provide sufficient knowledge to enable the readers to understand the underlying principles in wireless instrumentation and networks. Particular emphasis is given to the wireless communication systems since it is an important area of science and technology that is expanding very rapidly thus impacting all aspects of our daily lives. It is also one of the most promising areas of research and development.

Wireless communication technologies have become very popular and there are hundreds of wireless equipment manufacturers and perhaps as many standards. Understanding the benefits and shortfalls of each equipment and associated standards can make the selection and implementation easier. In this respect this book provides guidance to state-of-art of wireless technology as it offers many benefits in measurement applications. Some of the benefits can be lowering the wiring costs, simplifying data transfer, extending the distance of communication, enabling remote monitoring, and providing flexibility in the networking of the devices.

The purpose of *Wireless Sensors and Instruments: Networks, Design, and Applications* is to provide knowledge and guidance for engineers, scientists, designers, researchers, and students who are involved in measurement and instrumentation. This book covers concepts in instruments and instrumentation, electromagnetic wave propagation, radio engineering, digital networks, wireless sensors and instruments design and applications. Each chapter includes descriptive information for professionals, students, and technical staff involved in measurement and control. Numerous equations are given to assist engineers and scientists who wish to solve problems that arise in a fields other than their specialties.

Acknowledgments

Many people have directly and indirectly contributed to this book. I thank the people, colleagues and those whose paths I have crossed during my working life. My particular thanks are to Professor Kit Po Wong at Hong Kong Polytechnic University for providing encouragement to go ahead with this book while I was appointed visiting professor at his institution. Writing books is a lonely affair requiring major commitments and a great deal of patience and determination. The encouragement of Kit Po has been extremely valuable when I needed it most. I would like to thank my colleagues at Gazi University, Department of Electrical and Computer Engineering, Ankara, for providing an office and computer facilities in the initial and final stages of writing.

I would like to thank all the companies that provided information on their products and granted permission to reprint some of the images of their equipment. In particular my appreciations are extended to Steven Arms of Microstrain Inc.; Wayne Magnes of Oak Ridge National Laboratory; Graham Moss of Elprotech; Randy Culpepper of Texas Instruments; Colleen Cronin of Analog Devices, Inc.; Colin Pickard of Oregon Scientific; and Shana Jacob of CrossBow Inc.

Special thanks are extended to the CRC Press staff who made this book possible, especially to Nora Konopka for guiding the book through completion; also thanks to Helena Redshaw, Manager of Editorial Project development; and Jay Margolis, Project Editor, for their patient and professional involvements in putting it all together.

Author

Halit Eren, Ph.D., received the B.Eng. degree in 1973, the M.Eng. degree in electrical engineering in 1975 and the PhD degree in control engineering from the University of Sheffield UK. He earned an MBA from Curtin University of Technology, Western Australia in 1998 majoring in international management and leadership.

Dr. Eren has been lecturing at the Curtin University of Technology since 1983, first at the Kalgoorlie School of Mines and then School of Electrical and Computer Engineering, Perth, Western Australia. He served as the head of the Department of Electronic and Communication for some time.

Dr Eren's expertise lies in the area of instruments, instrumentation systems, and networking; electronic portable instruments, signal processing; and engineering mathematics. He has been researching in wireless and portable instruments for more than 17 years, mainly in the areas of electromagnetic, ultrasonic and infrared techniques, fieldbus, telemetry, and telecontrollers. He serves as a consultant to many industrial and government organizations. Dr. Eren has contributed to more than 150 conferences, journals and transactions and various books published by the CRC Press and John Wiley & Sons.

Dr Eren is the author of the book entitled *Electronic Portable Instruments: Design and Applications* published by the CRC Press in 2004. This second book on wireless sensors and instruments networks, design and applications forms a synergy with the first book as the application of modern technology is producing a large number of portable and wireless instruments.

Introduction

Instruments are essential for measuring physical variables in industrial operations, consumer products, environmental monitoring, research and development (R&D), transportation, military, space exploration, avionics, and so on. A collection of instruments form an instrumentation system, which may be responsible for numerous measurements in a complex process. Instruments are networked to communicate directly with each other or via intermediate devices such as computers, microprocessors, or controllers. Today, traditional instrumentation systems communicate through wired media. However, communication between instruments by wireless techniques is developing rapidly and gaining broad acceptance. It is very likely that wireless instruments will replace their wired counterparts in the very near future.

In recent years, considerable progress has been made in instruments and instrumentation systems due to the employment of integrated circuit technology, analog and digital components, efficient and low-power microprocessors, intelligent sensors, radio frequency (RF) communication technology, and protocols and standards supporting networks. In particular, cost-effective RF products have expanded suddenly to unimaginable dimensions. Devices such as cellular and cordless telephones, private and public telephone systems, wireless modems, radio frequency identification (RFID), and wireless sensors and instruments have rapidly penetrated into all aspects of our lives. Most of these devices were initially available as rare and expensive luxury items, but are now used by consumers, industry, and scientific and other communities. With this growth in demand, small and large semiconductor and system vendors are competing for a larger market share by introducing a diverse range of RF products.

Most people are familiar with a number of wireless control and communication systems used in everyday life. Mobile cellular telephones, cordless telephones, hand-held walkie-talkies, pagers, garage door openers, remote controllers, home entertainment equipment controllers are some of the examples that can be mentioned. The proliferation of wireless systems in consumer products and industrial applications is so fast that a progressive involvement in technological developments are vital for governments and industrial establishments alike if they want to stay competitively in the rapidly changing field of wireless communication systems and their applications.

Today's wireless networks were largely developed for mobile telephones and service mainly voice-based applications. Nevertheless, this is changing rapidly, with great emphasis being placed on wireless data transmission and wireless access to the Internet. For example, the aim of third-generation wireless systems, which provide access to the Internet, video, etc., is to make personal communication available anywhere at any time. Similarly, the application of wireless techniques to measurement and instrumentation is enabling configuration of wireless instrumentation networks in many applications, ranging from intelligent buildings to implantable wireless sensors used for human health improvement.

Wireless instrumentation requires a broad knowledge involving a diverse range of disciplines, including measurement science, design and construction of electronic circuits, understanding of integrated circuit (IC) technology, appreciation of the theory and practice of wireless communication systems, networks, and protocols and standards. In this book, all these concepts will be explained sufficiently to enable readers understand the underlying principles in wireless instrumentation and networks. Particular emphasis will be placed on wireless communication systems, since it is in this area that science and technology are expanding. Wireless communication systems are also a promising area of research, with thousands of researchers concentrating on the topic.

International standards are rapidly emerging in wireless technology as applied to sensors and instruments. Some of important ones are Bluetooth, HiperLAN, and the IEEE 802 standards for communication and networks, and the IEEE 1451 family of standards for sensors. These standards are gaining wide acceptance. Communication among devices involved can take place as point-to-point or point-to-multipoint. This gives considerable flexibility in network configuration and communication algorithms can be tailored to improve system reliability and adaptability.

This book is a reflection of information on the latest technologies in the field of wireless sensors, instruments, and networks. Engineers and scientists who are not trained in electrical engineering will find the book very informative without overwhelming them with detailed information.

This book was written for students and practicing scientists and engineers who are already familiar with technical concepts in electronics, probability theory, communication theory, basic electromagnetic theory, networks, and operational aspects of networks. However, the information given combines materials from many different disciplines, therefore it is highly unlikely that all readers will have the necessary basic knowledge for the topics covered. Therefore necessary concepts throughout the book are developed from principles to accommodate a wide range of readers from different backgrounds. This approach makes it attractive for practicing engineers and scientists who are involved in instruments and instrumentation in their disciplines. The book is also recommended as a useful teaching tool for undergraduate and postgraduate students who are likely to be involved with the design and operation of modern instruments and networks.

This book contains five chapters. The first three chapters provide information on measurements, instruments, sensors, communication systems, and networks. Information provided in these three chapters is brought together in Chapter 4 and Chapter 5.

In Chapter 1, brief but essential background information on measurements, instruments, instrumentation, sensor technology, communication systems, and networks is given. Wireless communication of instruments has been developing remarkably fast and is becoming common practice in industrial and many other applications. Wireless communication technology is able to address the needs of effective and efficient communication in all types of instruments and instrumentation systems. Digital instruments and their associated theory, methods, and components are highlighted. Since digital sensor technology forms the backbone of all types of modern instruments, extensive discussions are presented on this topic. Wireless instrument communication and its applications in industrial environments have been studied. It has been shown that noise, interference, and distortion can play a significant role in the operation of instruments and their associated networks.

Chapter 2 concentrates on modern communication systems. The chapter starts with the principles of electromagnetic wave propagation and expands to important characteristics of electromagnetic radiation, such as losses, fading, reflection, refraction, and attenuation. The necessary electronic components for successful RF communication are discussed. Applications of digital communication techniques are making wireless sensors, instruments, and networks possible, thus the fundamentals of digital communication technology are explained in detail. Modern communication methods, modulation, and multiplexing techniques, frequency spreading, and multiple access methods are also discussed and examples are given.

Chapter 3 discusses networks, protocols, standards, and topologies. Networks are collections of interoperational devices linked together by a communication medium and supported by suitable software. Networking of hardware and software resources is essential to bring multiple devices together to provide efficiency by enabling the exchange of information, creating collaborative operations, and sharing the functions of equipment and devices. In this chapter, the types of network topologies, protocols, and standards relevant to wireless networks are explained. The security of wireless networks is highlighted and the methods are discussed. The knowledge and experience gained in wired network technologies can be applied directly to wireless operations. Newly emerging wireless technologies such as the IEEE 802 family of standards and Bluetooth, among others, are discussed.

In Chapter 4, the construction of wireless sensors and instruments is introduced and examples are given. Instrument communication protocols are revisited and current technologies applied in wireless instruments and sensors are explained. Modern wireless sensor and instrument networks can produce using embedded or modular designs. These networks can be expanded using bridges, routers, and repeaters. The construction of wireless sensors and instruments is discussed and many examples are provided. Power issues of wireless networks are also addressed. Wireless sensor networks and wireless integrated sensor networks are detailed. Applications of Bluetooth and IEEE 802 technologies are demonstrated.

Chapter 5 is dedicated to the applications of wireless sensors and instruments—ranging from complex industrial plants to tracking wildlife in the wilderness. New application areas are being added as vendors of wireless sensors and instruments respond to consumer demands by offering a diverse range of wireless devices. In this chapter, examples of wireless sensors, instruments, and networks are provided in the following areas: specific applications, commercial applications, R&D, industrial applications, human health, environmental applications, RFID, consumer products, and other applications.

This book reflects the current state of the art in wireless sensors and instrument technology and provides guidance to students, researchers, practicing engineers, and scientists. I hope you enjoy it, as well as gain valuable knowledge from it, so that you can put this knowledge in use in the area of your interest.

Table of Contents

1

Instruments and Instrumentation

Instruments are developed for sensing and measuring physical variables that are essential in industrial operations, environmental applications, research and development (R&D), transportation, military equipment, and in our daily lives. Instrumentation systems are collections of instruments networked to communicate with each other directly or via some intermediate device such as computers or microprocessors. The majority of instrument communication systems have been based on wired media, but today wireless communication is developing rapidly and becoming common in industrial and many other applications. In this chapter, brief but essential background information is provided on measurements, instruments, instrumentation, sensor technology, communication systems, and networks.

If the behavior of the physical variable is known, its performance can be monitored and assessed. Applications of instruments range from laboratory conditions to arduous environments, such as inside nuclear reactors, to remote locations, such as satellite systems or spaceships. Manufacturers produce a wide range of instruments in order to meet diverse requirements. The majority of modern instruments have a great degree flexibility in their range of uses, types of displays, and methods of communication with other devices.

In instrument communication, information generated by a source is passed to a sink. The source converts the measured or sensed variable into electrical signals. Electrical signals are then processed and modified into communication signals that are passed through a communication channel in the form of useful information or a message. The received signals are then converted back into signals at the sink. Information can be transmitted through wired or wireless media by a variety of techniques.

In recent years, considerable progress has been made in measurements, instruments, and instrumentation systems because of the progress in integrated circuit (IC) technology, the availability of low cost analog and digital components, and efficient microprocessors. Consequently the performance of measuring and monitoring instruments has improved significantly because of the availability of on-line and off-line analysis, advanced signal processing techniques, and local and international standards. Today's wireless communication technology is able to address the needs of effective and efficient communication in all types of instruments and instrumentation systems.

In this chapter, measurement issues are introduced and instrument architecture explained. Digital instruments and their associated theory, methods, and components are highlighted. Sensor technology is the backbone of all types of instruments, therefore sensors will be explained in detail. A general introduction is provided on instrument communication as applied to common instruments as well as those used in industry. Instrument networks and their associated standards and protocols are discussed. Overall, this chapter concentrates on the source and associated issues such as noise, distortion, and interference during communication.

1.1 Measurements

Measurement is a process of gathering information from the physical world within agreed upon national and international standards and procedures. Measurements are carried out using manmade instruments that are designed and manufactured to perform specific functions. The functionality of an instrument is to maintain a prescribed relationship between the measured numerical values and the physical variable, or measurand, under investigation.

A typical instrument has many components, the sensors and transducers being the primary elements that respond to the physical variable and generate useful signals. A sensor is a physical entity that converts a physical variable into a processable electrical signal. Sensor signals in the majority of modern instruments are in electrical form or they are ultimately converted to an electrical form. This is because electrical signals are easy to process, display, store, and transmit. Similar to sensors, transducers also convert energy from one form to another between two physical systems.

Transducers are used in a wide variety of industries, including aerospace, automotive, biomedical, industrial control, manufacturing, and process control. As the demand for automation systems grows, the need for transducers increases. This demand is partially met by the development of smart sensors and transducers and by the integration of wireless technologies, making today's instruments and instrumentation systems much more flexible, cheaper, and easer to use.

Once converted to electrical form, the relation between the sensor signals and physical variations can be expressed in a transfer function, a mathematical model between the sensor signal and the physical variable. In a continuous system, the transfer function may be linear or nonlinear. For example, a linear relationship is expressed by the following equation:

$$y = a + bx, \qquad (1.1)$$

where y is the electric signal from the sensor, x is the physical stimulus, a is the intercept on the y-axis, which gives the output signal for a zero input, and b is the slope, also known as the sensitivity.

The output signal y represents the physical variable in amplitudes, frequencies, phases, or other properties of electrical signals, depending on the design and construction of a particular sensor and the nature of the variable. The nonlinear transfer functions can be logarithmic, exponential, or in other forms of mathematical functions. In many applications, a nonlinear sensor signal may be linearized over some limited ranges.

For a specific measurement, a wide range of sensors and transducers may be available. Selection of the correct sensors, transducers, and components to retrieve the required information and employment of representative signal processing techniques is essential.

1.2 Instrument Architecture and Instrumentation

Instruments are manmade devices that are designed and constructed using the existing knowledge about a physical process and the available technology. Appropriate hardware and software are engineered that can perform well within the expected specifications and standards.

The functionality of a typical instrument can be broken into smaller components, as illustrated in Figure 1.1. All measuring instruments have some or all of these components. Instruments differ from each other in the way they handle, transmit, and display information.

Generally signals from sensors are not suitable for displaying, recording, or transmitting in their raw form. The amplitudes, power levels, or bandwidths of sensor signals may be very small or may carry excessive noise and superimposed interference that masks the desired components. Signal conditioners adapt sensor signals to acceptable levels and shapes for processing and display.

1.2.1 Signals and Signal Conditioning

Measurement of physical variables generally leads to the generation electrical signals. A signal can be defined as "any physical quantity that varies with time, space, or any other independent variable or variables." Naturally, different types of signals require different processing techniques, hence different types of analog and digital methods must be used and associated components must be selected.

The analysis and processing of signals requires some kind of mathematical description of the signal. A signal can be described as a function of one or more independent variables. Therefore signals generated by sensors can be

FIGURE 1.1
Essential components of an instrument.

classified in a number of mathematical ways, such as multichannel and multidimensional signals, continuous time or discrete signals, deterministic or random (stationary, nonstationary) signals, transient signals, etc.

In some applications where multiple signal sources are present, multiple sensors are used to generate signals. These signals can be represented in vectors. Generally if the signal is a function of a single independent variable, it is called a one-dimensional signal. Similarly a signal is called *m*-dimensional if its value is a function of *m* independent variables (e.g., earthquake signals picked by various accelerometers). Vectors in matrix form can represent such signals, and appropriate techniques are applied for processing.

Any signal that can be expressed by an explicit mathematical expression or by well-defined rules is called a deterministic signal. That is, the past, present, and future value of the signal is known precisely without uncertainty. In this respect, the signal may be classified according to the characteristics of an independent variable (e.g., time) and the value it takes. Deterministic signals can be continuous or discrete.

Continuous signals, also known as analog signals, are defined for every value of time from $-\infty$ to $+\infty$. Continuous signals can be periodic or nonperiodic. In periodic signals, the signal repeats itself in an oscillatory manner, which can be represented as a sinusoidal waveform:

$$x(t) = X_m \sin(\omega t),\tag{1.2}$$

where $x(t)$ is the time-dependent signal, ω is the angular frequency (2π ft), and X_m is the maximum value.

Continuous signals can be periodic, but not necessarily sinusoidal, such as triangular, sawtooth, rectangular, or other regular or irregular shapes. If the signals are periodic but nonsinusoidal, they can be expressed in Fourier series as a combination of a number of pure sinusoidal waveforms as

$$x(t) = X_0 + X_1\sin(\omega_1 t + \phi_1) + X_2\sin(\omega_2 t + \phi_2) + \dots + X_n\sin(\omega_n t + \phi_n),\tag{1.3}$$

where $\omega_1, \omega_2, \dots, \omega_n$ are the frequencies (rad/s), $X_0, X_1, \dots, X_n$ are the maximum amplitudes of respective frequencies, and $\phi_1, \phi_2, \dots, \phi_n$ are the phase angles.

In Equation 1.3, the number of terms may be infinite, and the greater the number of elements the better the approximation. These elements constitute the frequency spectrum. The signals can be represented in the time domain or frequency domain, both of which are extremely useful in the analysis.

Discrete signals are defined only at discrete intervals of time. These time intervals may not be equal, but in practice, for computational convenience, they are assumed to be equally spaced. Equation 1.2 can be represented in discrete form as

$$x(nT) = X_m \sin(\omega nT), \qquad (1.4)$$

where

$$t = nT, \ n = 0, \pm1, \pm2, \pm3, \ldots.$$

Digital signals are represented as 1s and 0s. These signals can be generated digitally or obtained from analog signals by the application of appropriate analog-to-digital (A/D) signal converters. Digital signals have many advantages over analog signals. Since they are only 1s and 0s, they are easy to generate, process, store, multiplex, and transmit. They are relatively immune to noise, error corrections can be carried out easily, and encryption and other security issues can be easily addressed. However, they require greater bandwidths in communication, particularly if wireless techniques are used.

Random signals vary randomly in time. Random signals are often seen in nature, where they constitute irregular cycles that never repeat themselves exactly. Random signals cannot be deterministically described to any reasonable degree of accuracy or a description is too complicated for practical use. However, statistical methods and probability theory can be used for analysis by taking representative samples. Theoretically an infinitely long time period is necessary to obtain a complete description of these signals. Mathematical tools such as probability distributions, probability densities, frequency spectra, cross correlations, autocorrelations, digital Fourier transforms (DFTs), fast Fourier transforms (FFTs), and autospectral analysis, root mean square values, and digital filter analysis are some of the techniques that can be employed.

If the statistical properties of signals do not vary over time, the signal is called a stationary random signal. However, in many cases the statistical properties of signals vary over time, producing nonstationary random signals. In this case, methods such as time averaging and other statistical techniques can be employed.

Signals can also be classified according to their amplitude level, the relationship of their source terminals and ground, bandwidth, and also with respect to their output impedance properties. As far as amplitudes are concerned, generally signals lower than 100 mV are considered to be low level signals that need amplification.

Single-ended signals come from a source that has one of its two output terminals kept at a constant voltage. Usually one of the terminals is grounded and that terminal serves as a common point for other signals. A source of a differential signal, on the other hand, has two output terminals whose voltages change simultaneously by the same magnitude, but in opposite directions. Any other differential signal from the same source must have two other differential terminals. The output voltage of a bridge is a typical example of a differential signal.

Narrowband signals have very small frequency ranges relative to their central frequencies. A narrowband signal can be at a direct current (DC) level, or almost DC, as a result of low-frequency signal sources (such as thermocouples), or at alternating current (AC) levels, such as those obtained from AC-driven modulating sensors.

Broadband signals have a large frequency range relative to their central frequency. In broadband signals, the value of the central frequency is important; for example, a signal from 1 Hz to 10 Hz may be considered to be a broadband instrumentation signal, whereas two 20 kHz sideband signals around 2 MHz may be regarded as a narrowband signal.

1.2.2 Types of Instruments

Instruments can broadly be classified as analog instruments, or digital instruments, or the combination of the two. Most modern instruments are digital because of many advantages that they offer. Some of the advantages manifest themselves in their flexibility in design, programmability, ease of use, convenience in communication with other devices, etc. However, from the manufacturers' and designers' point of view, the front ends of many instruments are still analog, since majority of sensors and transducers generate analog signals. Analog signals are converted to digital forms for digital instruments and instrumentation.

Instruments can also be classified as portable or nonportable (fixed) instruments. Portable instruments rely on an onboard power source and can be used in different places; this mobility provides many advantages. Nonportable instruments have unlimited power from their power source and are extensively used in laboratory and industrial environments. Fixed instruments are part of a system, such as the dashboard instruments in vehicles and airplanes, and the machinery in industry, buildings, power stations, etc. Although instruments can be portable or nonportable, they all operate on analog or digital principles.

Analog instruments operate purely on analog principles, thus they measure, transmit, display, and store data in an analog form. Analog instruments make use of continuous changes in current and/or voltage amplitudes, phase differences, or frequencies, or combinations of some or all of these. Signals from the physical variables that analog instruments need to handle can be deterministic or nondeterministic, and may contain various forms of extensive noise. The signal conditioning in analog instruments is usually realized by integrating many functional blocks such as bridges, amplifiers, filters, oscillators, modulators, offset circuits, level converters, and buffers. Some of these functional blocks are illustrated in Figure 1.2.

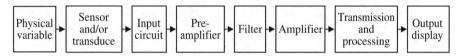

FIGURE 1.2
A detailed block diagram of an analog instrument.

One of the essential building blocks of analog instruments is operational amplifiers (op amps) and instrumentation amplifiers. Operational amplifiers are integrated devices that are made from either monolithic or hybrid components. An operational amplifier may contain hundreds of transistors, resistors, and capacitors in a single chip. They can be configured as inverting or noninverting amplifiers. By means of appropriate external components, they can perform many other functions such as multipliers, adders, limiters, integrators, differentiators, filters, etc.

Digital instruments largely operate on digital principles. In many cases, the signals generated by sensors and transducers are conditioned first by analog circuits before they are converted into digital signals for further processing. However, digital instruments operating on purely digital principles are now being produced thanks to micro- and nanotechnologies in the construction of modern sensors and supporting devices. For example, today's smart sensors based on digital electronics contain the complete signal condition circuit in a single chip integrated with the sensor itself. Thus the output of many smart sensors can be interfaced directly with other digital devices.

Microinstruments, also known as microsensors, are a typical example of modern digital instruments, as exemplified in Figure 1.3. This single-chip

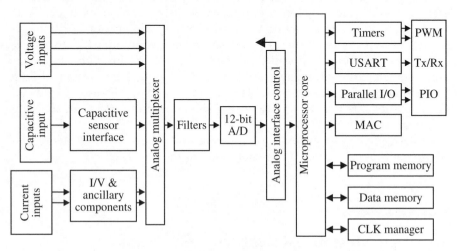

FIGURE 1.3
A block diagram of a typical microinstrument.

implementation of a microinstrumentation system is based on complementary metal oxide semiconductor (CMOS) technology (Kraver et al., 2000). This chip incorporates a voltage, current, and capacitive sensor interface; a temperature sensor; a 10-channel 12-bit A/D converter; and an 8-bit microcontroller with a 16-bit hardware multiplier and 40-bit accumulator. This device operates on a 3 V power supply drawing 16 mA when fully powered or 850 µA at standby.

Digital communication (in the system in Figure 1.3) is facilitated by onboard serial and parallel ports. The system contains two parallel input/output (PIO) units and a universal synchronous/asynchronous receiver/transmitter (USART). The program memory consists of a 512 B boot read-only memory (ROM), 4 KB of random access memory (RAM) for general purpose use, and 512 B RAM for data storage. The multiple accumulate (MAC) unit supports 6-bit accumulation and 40-bit accumulation, allowing on-chip signal processing. Standard timing is provided by watchdog timers and multifunction timers.

The microinstrument illustrated in Figure 1.3 is supported by C programming language. It includes hardware support for a single break point and trace event for code debugging. A development system interface provides instructions for halting the processor, single-stepping through code, and reading and writing system registers. Single-cycle interrupt response and direct memory access (DMA) allows application of this device in time-critical applications. Clock manager is used to divide the system clock to reduce power, assign a slow clock, or halt the clock until a programmed event occurs.

The advantages of such microinstruments are that digital circuits, compared to their analog counterparts, are cheaper, more reliable, and more flexible. Since digital hardware allows programmable operations, it is possible to modify the functions of an instrument through its software. Thus digital hardware and its associated software provide greater flexibility in system design compared to an equivalent analog counterpart. However, it is worth mentioning that one disadvantage of digital instruments is that some signals with extremely wide bandwidths require intensive real-time processing requiring large amounts of memory and sophisticated software considerations. For such signals, analog signal processing may be the only solution.

Modern wireless instruments are almost exclusively digital, thus digital instruments are explained in greater detail in the next section.

1.3   Digital Instrument Hardware and Software

The signals acquired for physical variables are usually in analog form, therefore traditional digital instruments involve two distinct processing techniques: analog and digital. In the analog section, signals are processed in analog and converted into their digital counterparts before they are passed on to the next stage of the system for further processing. In the conversion process, multiplexers, sample-and-hold (S/H) devices, and A/D converters are used. In recent years, digital instruments have been manufactured in the form of sophisticated microinstruments, as illustrated in Figure 1.3. Micro-instruments integrate all the essential components of an instrument in a single chip. The typical components of a traditional digital instrument are illustrated in block diagram form in Figure 1.4.

A typical digital instrument consists of five main subsystems: the analog front end for signal generation and conditioning, the general purpose pro-grammable digital hardware, storage and communication components, application-specific input/output (I/O) components, and other ancillaries such as displays and power supplies. All of these components may not be present in some of the simpler types, depending on their functions in a system. However, in all digital or semidigital instruments, the sensor signals are amplified or attenuated, filtered by analog signal conditioning circuitry, and converted to their digital representations by an A/D converter. The digital signals are then processed in accordance with the measurement algo-rithms and the results are displayed, stored, or transmitted to other devices for further processing or storage. Signals generated by some sensors may be sufficiently high in quality to be interfaced directly to a digital system, as in the case of various optical systems used for motion and rotation sensing.

Once the signals are converted into digital forms, the data can be processed by employing various techniques such as FFT analysis, digital filtering, sequential or logical decision making, correlation methods, spectrum anal-ysis, etc.

Modern digital instruments are particularly useful in performing complex operations, storing and transmitting information, and communicating the relevant data. This is because these instruments are based on digital proces-sors, which provide powerful and flexible signal processing and data han-dling capabilities. Signal processing and data handling can be achieved simply by software or firmware implementations. Another important advan-tage of digital instruments is that once the signals are converted into digital form, they can be managed, shared, and processed by other types of digital devices such as computers and microprocessor-based systems. This provides a wide range of possibilities for data processing, communications, storage, and visual displays.

Figure 1.5 is a block diagram of a typical microprocessor-based digital instru-ment. Many of the microprocessor-based instruments available today can

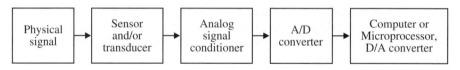

FIGURE 1.4
Components of a typical digital instrument.

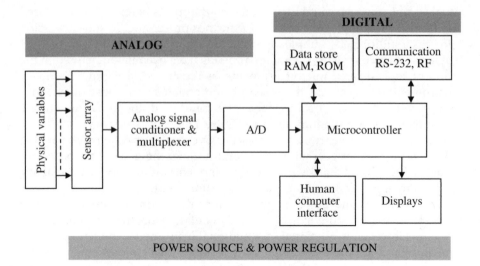

FIGURE 1.5
A typical microprocessor-based portable instrument.

implement complex and sophisticated measurements, store large numbers of measurement results, display the information graphically, and can function as part of a complex measurement system. Most of these devices can operate in an independent manner as stand-alone instruments or use advanced communication techniques to integrate as part of a complex system.

Many modern digital instruments are equipped with effective communication capabilities, which are achieved in wired and wireless forms. In these instruments, integration of fast central processing units (CPUs), large memories, and advanced wireless communication make them flexible, easy to use, multifunctional, and networkable.

1.3.1 Components of Digital Instruments

The key components of digital instruments are the S/H device and the A/D converter. These components may limit instrument accuracy (especially in terms of linearity) and bandwidth. Other key components are microprocessors and microcontrollers, memory chips, I/O devices, signal converters, and digital signal processors.

A designer can select components for digital instruments from a wide range of signal processors, memories, peripherals, and other devices with varying degrees of performance, power consumption, and size. Similarly, recent technological developments in analog components have contributed to further improvements in operational bandwidths, sensitivity, and accuracy. Fast and sensitive A/D and digital-to-analog (D/A) converters are fairly inexpensive. The introduction of novel converter architectures that incorporate internal digital antialiasing filters and sigma-delta converters allow designers to shrink the front-end analog hardware to a minimum size. Digital instrumentation has also benefited from the general development of user-friendly operating systems, high-productivity software development tools, and so on.

Sample-and-hold devices and A/D converters are discussed in Section 1.3.4. Microprocessors, microcontrollers, memory, I/O devices, and digital signal processors are discussed next.

1.3.2 Microprocessors and Microcontrollers

Microprocessors and microcontrollers are at the heart of almost all types of modern digital instruments and instrumentation systems. They play an important role in data acquisition, data processing and control, and instrument communication. However, the operation of all digital instruments is governed by common mathematical theories, such as sampling theory and digital signal conditioning theory. The whole process of carrying out sampling, conversion, and signal conditioning is referred to as data acquisition. In this section, the basic functions of data acquisition systems and the supporting theoretical foundations are discussed.

Microprocessors and microcontrollers used in digital instruments and instrumentation perform the following basic functions:

- Data handling, including data acquisition, data processing, information extraction, data compression, interpretation, recording, and storage.
- Instrument control, including sensors, actuators, system resources, and control of other internal and external devices.
- Human-machine interface, which provides a meaningful and flexible user interface to the operators for information display and control.
- Procedural development, including lookup tables, diagnostics, calibration, etc.
- Communication, including wireless techniques.

Microprocessors are ICs that handle and process data in binary form. A general purpose microprocessor is used as the CPU in computers. The most common manufacturers of microprocessors are Intel, Motorola, AMD, Philips, Zilog, Atmel, and Hitachi, among many others. The most commonly used microprocessors in instruments and instrumentation systems can be categorized into three basic types: embedded controllers, digital signal processors, and dedicated computers.

Microcontrollers are special microprocessors that have built-in memory and interface circuits within the same IC. Because of their smaller sizes, simplicity, low costs, and power efficiency, microcontrollers are commonly used in instruments and instrumentation systems. Typical families of such microcontrollers are the Motorola MC68HC11 and the peripheral interface controller (PIC). There are many microcontrollers that are available today that are suitable for instruments and instrumentation systems and every day microcontrollers are introduced with new and improved features and functionalities.

A microcontroller can be defined as a single-chip device that has memory for storing information and is able to control read/write functions and manipulate data. The performance is determined by its size (i.e., the number of bits that the CPU can handle at one time). There are many types of microcontrollers available, offered by a wide range of manufacturers.

Many microcontrollers have similar components for data processing and handling. The typical architecture of a microcontroller is illustrated in Figure 1.6. The main components of a microcontroller perform the following functions:

- CPU—executes instructions that are specially written to perform certain tasks.
- Memory to store information. There are different memory devices:
  - ROM—stores permanent information and application programs.
  - RAM—stores data that are subject to continuous change during the execution of the program.
  - Erasable memory, such as erasable programmable read-only memory (EPROM) and electrically erasable programmable read-only memory (EEPROM), to store mainly application programs, but which is erasable and reprogrammable.
- Interface components—communicate with external devices and components.

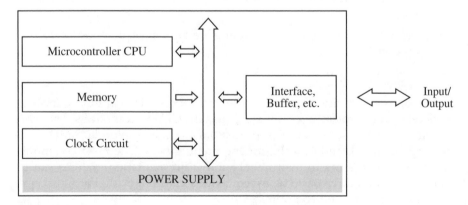

FIGURE 1.6
A block diagram of a microcontroller.

Two examples of microcontrollers that are often used in instruments and instrumentation systems are the Hitachi H8 and Motorola HCS08 microcontrollers. Further examples of microcontrollers are provided in Chapter 4 and Chapter 5.

Hitachi H8/300H

The H8/300H Tiny series is a group of 16-bit devices based on the H8/300H 16-bit CPU core, with different memory sizes and peripheral features. This group of microcontrollers have the same pinout so they can be interchanged between devices. The 16-bit microcontroller has sixteen 16-bit general registers and 62 basic instructions. It has four timers:

- Timer A—an output pin for divided clocks.
- Timer V (8 bits)—an output pin for waveforms generated by output compare.
- Timer W (16 bits)—functions include output compare output, input capture input, and pulse width modulation (PWM) output.
- Watchdog timer.

They use Hitachi Embedded Workshop assemblers/compilers. Other characteristics of the H8/300H microcontroller include 29 general I/O pins, 8 input pins for A/D converters, EEPROM interface, inter-integrated circuit (I^2C) bus interface, serial communication interface (SCI), and support for various power-down states.

Motorola HCS08

The Motorola HCS08 microcontroller is a typical example of low-power microcontrollers used in embedded instruments. HCS08 microcontrollers are low-power 1.8 V, 8-bit devices with an on-chip debug capability. They can be extended to operate as 16-bit microcontrollers. They were developed primarily for applications such as universal remote control units that are idle for long periods of time, but must power up quickly on demand. Other applications include handheld instruments, utility meters, security systems, and other portable and wireless industrial and consumer devices. The features of the HCS08 include

- Multiple power management modes, including a 20 nA power-down mode.
- A zero-component auto-wakeup from "stop" to 0.7 μA.
- Up to a 40 MHz CPU/20 MHz bus at 2.1 V and 16 MHz CPU/8 MHz bus at 1.8 V.

- A programmable internal clock generator with temperature and voltage compensation (typical drift less than 2%) for reliable communication and fast startup.
- In-application reprogramming and data storage via third-generation 0.25 μm flash technology.
- High integration, including four serial communication ports, up to 8 timer/PWMs, and an 8-channel 10-bit A/D converter specified down to 1.8 V.

An important feature of digital instruments is their ability to communicate with other devices using a wide range of communication methods. The internal and external communication of digital instruments is accomplished by the I/O components, discussed in the next section.

1.3.3 Input and Output

Data communication between microcontrollers and microprocessors is accomplished through the I/O devices. A variety of I/O hardware and software is available, depending on where the communication is taking place and the types of devices involved. Some of the I/O devices used in instruments and instrumentation systems are explained below.

- Universal asynchronous receiver transmitter (UART)—a port adapter for asynchronous serial communication.
- Universal synchronous/asynchronous receiver transmitter (USART) —a serial port adapter for either asynchronous or synchronous serial communication.
- Synchronous serial port—do not require start/stop bits and can operate at much higher clock rates than asynchronous serial ports. They are used for communicating with high-speed devices such as memory servers, display drivers, additional A/D ports, etc. In addition, they are suitable for simple microcontroller networks.
- Serial peripheral interface (SPI)—a type of synchronous serial port. Another version is the serial communication interface (SCI), an enhanced form of a UART.
- I²C bus—a simple two-wire serial interface developed by Philips. It was primarily developed for 8-bit applications and is widely used in consumer electronics, automotive, and industrial applications. The I²C bus has two lines and is useful for multimaster and multislave network interfaces with collision detection. Up to 128 devices can be connected on the network and they can be spread out over distances of 10 m.

- Controller area network (CAN)—a multiplexed wiring scheme that was developed jointly by Bosch and Intel for wiring in automobiles. The CAN specification is being used in industrial control both in North America and Europe.

Recent technological progress in wireless communication systems is having a significant impact on the development and use of all types of digital instruments. Wireless instruments can transmit the data acquired to a dedicated base station or can communicate among themselves, forming local networks. The desired signals acquired by sensors and transducers are encoded into signals suitable for wireless transmission via transceivers. On the receiving side, the signals are then demodulated and useful signals are extracted by suitable methods. In many applications, many different signals are multiplexed and sent via the same link. The transmission link can be realized by very high frequency (VHF) and other radio waves, microwaves, infrared, laser, etc.

1.3.4 Signal Conversion

In the majority of instruments, analog signals generated by sensors and transducers have to be conditioned in analog form before they are converted into their digital equivalents. Analog signal conditioning has two main goals: (1) the amplification of small signals or the attenuation of large ones, and (2) filtering for the removal of unwanted frequencies by suitable circuits.

Amplification or attenuation is used to adjust the signal amplitude range to the range required by the A/D converter. This is done because the voltages generated by sensors are generally much smaller than those that can be managed by A/D converters. In addition, the inputs of most A/D converters are unipolar; that is, one of the terminals of the converter is connected to a reference terminal. If the signal provided by the sensor is differential, the amplifier must convert this signal into unipolar form or a differential amplifier with a single-ended output must be utilized.

Filtering is used to modify the frequency spectrum of signals generated by sensors. There are two reasons for using filters: (1) to reduce interference, and (2) to narrow the bandwidth of the signals. Narrowing the bandwidth reduces the noise in the signal outside the band of interest and avoids possible aliasing.

Depending on the nature of the signal, other signal conditioning techniques may be applied, including adders, multipliers, limiters, etc. Once analog signal conditioning is complete, signals can be converted into digital form. Digitization of an analog signal involves sampling of the signal in time, followed by conversion of the sampled signal into its digital equivalent.

During the conversion stage, the A/D converter quantizes the sampled analog signal to one of the finite set of discrete numeric values available in the converter. The sampled analog signal is thus transformed into a digital form. Once created, digital signals are just arrays of numbers stored in memory and their association with time (sampling rate at which they were acquired) is stored and maintained by the signal processor independently of incoming signals.

Sampling and conversion of signals is the first stage of digital signal processing. One fundamental requirement of A/D conversions is that the digital signal obtained in a conversion process must be fully representative of the original analog signal. This simply means that the analog signal should be completely recoverable, through the reconstruction process, using the values of the digital signal. To guarantee the complete recoverability of an analog signal, the sampling period used to carry out the conversion must fulfill the postulates of the sampling theorem.

The sampling theorem states that "the number of samples per second must be at least twice the highest frequency present in the continuous signal." As a rule of thumb, depending on the significance of the high frequencies, the sampling must be about 5 to 10 times of the highest frequency component in the signal. This theorem, which is of crucial importance in the theory of signal acquisition, ensures that "under certain conditions a continuous time signal can be represented completely by and recoverable from the knowledge of its instantaneous values of samples equally spaced in time."

Analog-to-digital conversion involves four distinct stages: multiplexing, sampling, conversion, and quantization and coding. Multiplexing is necessary if there is more than one signal presented to the converter. During the conversion process, three main types of devices are used: multiplexers, S/H devices, and A/D converters.

Multiplexing is carried out by multiplexers, which act as a series of switches that select a particular signal from a multisensory system to be converted into digital form. The functionality of a multiplexer is illustrated in Figure 1.7. The multiplexing process can be carried out by two methods: analog or digital multiplexing. A multiplexer connects the analog channel to the S/H circuit. The S/H circuit samples the signal and holds it so that the A/D converter can carry out the conversion without errors. Once the

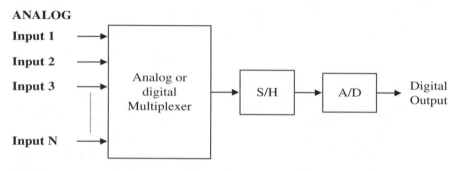

FIGURE 1.7
Functionality of a multiplexer.

conversion takes place, the output of the converter is connected to a common bus and data that belong to a particular channel are handled by the software.

Irrespective of the type of multiplexing selected, multiplexers are usually under the control of a central processor, microcontroller, or computer. This approach allows for the use of small-size devices, with low power consumption and cost savings. However, multiplexing has a number of problems. For instance, multiplexing generates switching transients at the input of the A/D converter, so it is necessary to allow the signals to settle before conversion takes place. Transients can propagate through the entire chain of amplifiers and filters. Another undesirable effect of multiplexing is signal leakage (cross talk) between channels due to switching effects and stray capacitances in the multiplexer. These effects may drastically limit the rate of data acquisition. Finally, if each data acquisition channel requires different settings and parameters, such as gain, frequency response, etc., then these parameters must be switched on a per-channel basis, which can be difficult, slow, and expensive.

Sampling is realized by S/H devices. They compensate for shortfalls in A/D converters due to conversion times. Usually the input signals to the A/D converter change over time due to changes in physical phenomena. If the input signal changes during the conversion process, uncertainties are introduced in the digital output. Full conversion accuracy can only be obtained if this uncertainty is kept below the conversion time, which is done by S/H devices.

Figure 1.8 shows the basic structure of a S/H circuit. The circuit contains a capacitor, a switch with its corresponding control circuitry, and two amplifiers to match the input and output impedances of the external circuits. The control circuit sends the commands SAMPLE and HOLD alternately to the switch S_1. This switch closes when it receives the SAMPLE command. Then capacitor C_h charges up (or down) until it reaches the level of the input signal, and a sample is taken. The control circuit immediately generates a HOLD command, which causes switch S_1 to open. Then the sampled voltage is held across capacitor C_h until a new sampling operation begins.

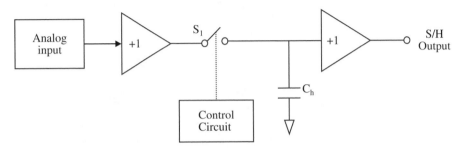

FIGURE 1.8
Basic structure of a S/H circuit.

Mathematically, this means

$$\left(\frac{dV}{dt}\right)_{max} \leq \frac{FS}{2^n t_c} , \qquad (1.5)$$

where n is the number of bits of the converter, t_c is the conversion time, and FS is the full-scale value.

This formula indicates that full conversion accuracy can never be obtained. That is the reason why S/H circuits are utilized. The S/H circuit acquires the value of the input analog signal and holds it throughout the conversion process until the conversion is complete, resulting in a zero drift in voltage. Once a fixed level of voltage is provided to the converter during the conversion process, the uncertainty caused by time changes in the input voltage is eliminated.

In digital instruments, sampling is usually carried out at equidistant time instants. However, when sampling is carried out at equidistant time instants, some conflict may arise between bandwidth and accuracy. In the case of high accuracy, A/D converters require conversion times that strongly limit the maximum sampling frequency and bandwidth if sampling instants are equally spaced. In order to overcome this limit, different strategies for asynchronous random sampling can be applied. Through suitable digital filtering over sufficiently long times, it is possible to use pseudorandom sampling to realize instruments whose bandwidth is not limited by A/D conversion times.

Conversion is realized by A/D converters and D/A converters. A/D and D/A converters provide an interface between analog and digital environments for computation and data processing purposes. Thus they form a vital link between real-time happenings and the digital environment.

All modern A/D converters produce digital output values in binary code, that is, the digits assume a value of either one or zero. The length of the output word defines the theoretical resolution of the A/D converter and the range of digital values that an A/D converter can produce. The smallest change in the input voltage that an A/D converter can detect is

$$V_{LSB} = \frac{V_{max} - V_{min}}{2^n} , \qquad (1.6)$$

where n is the number of bits produced by the converter and V_{max} and V_{min} are the maximum and minimum input voltages that an A/D converter can process correctly.

There are many different types of A/D converters available commercially. The most popular A/D converters are the counter-ramp, successive approximation (serial), flash (parallel), and sigma-delta converters. These converters are briefly discussed below. More information on A/D and D/A converters can be found in most instrumentation books (e.g., Eren, 2004).

Counter-ramp converters are simple devices and are slowest in the A/D converter family. An improvement to the counter-ramp A/D converter is the tracking converter.

Successive approximation converters operate with comparators. The conversion speeds of successive approximation converters are much greater than counter-ramp converters, but there is a trade-off between speed and complexity. These converters require extensive programmer logic, which takes the D/A decoder through different steps of the successive approximation process.

Flash or parallel converters are fastest, but are also the most complicated A/D converters. A flash converter, like the successive approximation converter, works by comparing the analog input signal with a reference signal, but a flash converter has as many comparators as there are digital representation levels or digital word length except for one. Therefore $2^{n-}1$ comparators are needed for an n-bit converter. The conversion rate for flash A/D converters is very fast because only one step is required for complete conversion. Depending on the propagation time of the encoding logic, digital output can be generated almost in real time.

The main disadvantage of flash converters is the number of comparators required to implement large-scale converters. For every binary bit of resolution added to an A/D converter, the number of comparators required is essentially doubled. For instance, 255 comparators are required for an 8-bit converter. If 4 more bits are added, for a 12-bit converter, 4095 comparators are required.

Sigma-delta converters, sometimes referred to as delta-sigma converters, are based on a very different method that limits analog processing to a few simple essential steps and uses digital signal processing techniques to achieve the required performance. This approach results in very high accuracy (up to 24 bit) at the low frequency range (1 kHz to 10 kHz) and moderate to high resolution (12 bits to 16 bits) in the mid-frequency range (10 kHz to 5 MHz).

Sigma-delta converters tend to achieve a higher degree of integration into the circuits and are smaller in size. Another feature of most sigma-delta converters is that their entire system can be scaled in frequency (including internal filters) simply by changing the frequency of the external sampling clock. At low sampling rates, sigma-delta converters can achieve very high resolutions (20 bit and even 24 bit). This simplifies data acquisition for small, low-level, low-frequency signals such as strain gauges, thermocouples, and so on, by allowing direct connection of the sensor and the A/D converter.

Quantization and coding are obtained as a result of A/D conversion when analog signals are represented in their binary equivalent forms. Amplitude quantization is defined as the process by which a signal with a continuous range of amplitudes is transformed into a signal with discrete amplitudes. The quantization process used in practice is usually memoryless. That is, the quantizer does not depend on previous sample values to predict the next value. While this is not an optimal solution, it is commonly used in practice.

Coding represents the analog signals in the digital domain as sequences of n bits. A different sequence of bits is assigned to each output state of the quantizer for each sample. For n bits, 2^n different sequences can be generated and consequently 2^n output states can be coded. The codes formed using bits are normally known as binary codes, and they can be of two types, unipolar and bipolar.

From the signal processing point of view, an A/D converter can be described in terms of the noise it generates. Quantization, like sampling, introduces some degree of uncertainty in the value of the analog signal before quantization takes place. For A/D converters with large numbers of bits (8 bits or more), this effect can be modeled as additive noise. The amount of quantization noise relative to the signal, or the signal:noise ratio depends on the nature and magnitude of the input signal.

In addition to quantization noise, A/D converters have other sources of noise, including internal voltage reference noise and thermal amplifier noise, among others. In practice, most converters rarely achieve their theoretical noise floor. Therefore, when selecting an A/D converter, one uses the number of bits of the converter as a rough estimate of its resolution. Nevertheless, close examination of the comprehensive converter specifications should reveal its true performance, and this may be necessary in some critical applications.

Once the signals are converted and represented in code, many other digital signal processing techniques can be applied, such as filtering, time shifting, spectrum analysis, and so on. Some digital signal processing techniques are discussed in the next section.

1.3.5   Digital Signal Processing

Digital signals are arrays of numbers that are used in the computational analysis of systems and signals. While analog signals are continuous in time and magnitude, sampled analog signals are discrete in time and continuous in magnitude. Digital signals are discrete both in time and magnitude. Digital signals can be generated by software or derived directly from analog signals using A/D converters.

Mathematically, signals are represented as functions of one or more independent variables. For example, time-varying voltage and current signals obtained from many sensors are functions of a single variable (i.e., time); the electromagnetic signals are functions of time and space in x, y, and z in Cartesian coordinates. Often signals from sensors need to be analyzed and processed to extract useful information. Fourier transforms and Fourier series are two common methods that can be used in signal analysis; both can be done in either the analog or digital domain. In these methods, the signal representation involves decomposition in terms of sinusoidal or complex exponential components. The decomposition is then said to be in the frequency domain. The decomposition is determined by frequency analysis tools that provide a mathematical and pictorial representation of frequency components, which is referred to as the frequency spectrum. The frequency spectrum can be determined by frequency or spectral analysis techniques. The Fourier series is useful in representing continuous-time periodic signals, whereas Fourier transforms are useful in representing periodic as well as nonperiodic signals, which are also known as finite energy signals.

In some cases, digital signal processing involves transformation of independent variables. Some important operations are time shifting, reflecting, and time scaling of signals. Time shifting of signals is often applied in radar, sonar, communication systems, and seismic signal processing. Reflection of signals is useful in examining symmetry properties that a signal may possess. Time scaling is useful to observe the behavior of a signal in expanded or contracted time to observe required details.

Digital filtering is one of the most commonly applied digital signal processing techniques. It is used to remove unwanted signal components, such as distortion or noise, or alternatively, to select only the desired components such as a particular frequency band. The two most common digital filters are the finite impulse response (FIR) filter and the infinite impulse response (IIR) filter. Both of these filters can easily be implemented by microprocessor systems and digital signal processors.

Digital signal generation is used extensively in many applications. Examples include mobile communication, sensor excitation, and testing and performance measurements of complex signal processing devices and systems, among others. Signal generation is used in the component testing measurements of inductive and capacitive parameters, in eddy current instrumentation, and as a local oscillator in modulation and demodulation processing.

1.4 Sensor Technology and Advanced Sensors

Sensors and transducers are the primary elements of instruments and instrumentation systems. They produce output signals in response to physical variations. Signals generated by sensors are processed suitably by analog or digital techniques to extract the desired information representing the behavior of the physical variations. Sensors are produced by a variety of techniques and from a wide range of materials.

1.4.1 Sensing Materials

Sensors are fabricated from materials such as metals, polymers and plastics, ceramics and glass, biological materials, semiconductors (largely silicon), and combinations of two or more of these materials.

Metals are commonly used for manufacturing sensors. As sensing elements, metals exhibit many different properties and characteristics. They can be electrically conductive, nonconductive, or semiconductive; magnetically ferrous or nonferrous; thermally conductive or nonconductive; mechanically elastic, brittle, strong, lustrous, or opaque; chemically a catalyst, chemiabsorbent, reactive, or nonreactive, and so on. Most of these properties are utilized in sensor technology, depending on the requirements of the physical variations. Some of the most common metals used in sensor technology are aluminum, beryllium, copper, lead, platinum, gold, and zinc. Aluminum exhibits good electrical, electromagnetic, and optical properties. Beryllium is used in optical and X-ray measurements. Magnesium, platinum, and nickel are preferred as the basic elements in temperature sensing. Copper and gold have good electrical and thermal properties and are used primarily in semiconductors and microsensors. Lead has good piezoelectric, sound, and vibration properties. Zinc is used as one of the main additives to improve the sensing properties of some alloys.

As well as primary sensing elements, metal alloys in various combinations are also used in sensor technology, particularly in temperature sensing. A good example of such alloys is chromel (90% nickel, 10% chromium). Metals and their alloys exhibit strong dielectric and polarization properties. Examples of some of their uses include electromagnetic and nuclear radioactivity measurements, and chemical and biological measurements. Metal alloys are also used as piezoelectric, pyroelectric, electrolyte, and ferroelectric materials responding to mechanical stress and strain, light, chemical stimuli, and magnetism, respectively.

Polymers and plastics have many applications in sensor technology. They are made from monomers, which react with each other to form long chains of polymers and polystyrenes. The primary element of polymers and plastics is carbon. Carbon atoms combine with other elements to produce thousands of different types of materials. Apart from carbon, only seven other elements are used to produce many forms of polymers and plastics. These elements are hydrogen, nitrogen, oxygen, fluorine, silicon, sulfur, and chlorine. By suitably combining these elements, large numbers of molecules can be produced that exhibit many different properties suitable for sensors. Plastics have many applications in sensors. For example, alkyd has good electrical as well as moisture absorption properties and epoxy has electrical properties. Plastics exhibit other properties, such as piezoelectricity, that have applications in stress, strain, vibration, and sound measurements.

Ceramics and glasses are useful materials for producing a range of sensors. Ceramics are made from several forms of metal carbides and nitrates, which exhibit thermal stability, are lightweight, are resistant to chemicals, and have structural strength and good electrical properties. For instance, silicon carbide has a high dielectric constant and is used in capacitive sensors. Ceramics are also used as basic elements in microsensor technology.

Glasses are based on silicate systems and are made from three major components: silica (SiO), lime ($CaCO_3$), and sodium carbonate ($NaCO_3$).

Glasses are produced in many different forms and shapes to give different optical properties, particularly in the mid- and far-infrared, visible, and ultraviolet (UV) ranges. Therefore glasses are used extensively in light-sensing applications.

Biological materials are mainly used for sensing chemical and biochemical species. Most biosensors consist of two parts, the biologically active component and the transducer component. The biologically active component provides a specific interaction with the target molecules. This interaction results in the release of some form of energy that can be sensed by the transducer component. Biosensors are manufactured and used for their different physical and chemical properties, such as antibody-based biosensors, enzyme-based biosensors, biological microresistors, etc.

Semiconductor materials are manufactured from silicon or germanium and are used extensively in modern sensor technology. Silicon is the most common material used in sensor technology, particularly in microsensors, intelligent sensors, and system-on-chip devices. Silicon exhibits favorable mechanical, chemical, and electrical properties. It is a relatively inert element that can only be attacked by a handful chemical substances such as halogens, dilute alkali, and hydrofluoric acid.

The properties of silicon and its applications in sensor technology are well researched. Silicon has a number of properties that make it useful in radiation, temperature, and optical measurements, as well as in mechanical, electrical, chemical, and magnetic measurements. For example, in radiation and optical measurements, silicon exhibits photovoltaic, photoelectric, photoconductive, and photomagnetoelectric properties. In temperature measurements, it exhibits the Seebeck effect, the Nernst effect, and temperature dependence of pn junctions. In chemical measurements, silicon exhibits ion sensitivity to certain chemicals by changing resistance currents, voltages, and other properties. It also has various magnetic properties such as the Hall effect, magnetoresistance, and the Suhi effect.

Silicon has good mechanical properties that are widely used to fabricate pressure transducers, temperature sensors, and force and tactile sensors. Thin film and photolithographic technologies allow silicon to be fabricated as small, high-precision mechanical structures suitable for microsensor technology. Micromachining allows the use of semiconductors to form microstructures suitable for sensing acceleration, force, pressure, mechanical impulses, etc. In other types of mechanical variables sensing, silicon has properties such as piezoresistivity, piezoelectric, capacitive effects, resonance, thermoelectric effects, etc.

One major problem with silicon is temperature dependence. Many of the properties in response to physical variables such as strain, light, and magnetic fields tend to be nonlinear and temperature dependent. The physical properties of silicon can be enhanced considerably by depositing different elements and materials in the crystal structure. For example, the addition of zinc oxide (ZnO) thin films into a silicon structure enhances piezoelectric properties and can be used in producing accelerometers and surface acoustic wave (SAW) filters.

Apart from oxygen, silicon is the second most abundant material on the Earth's crust, existing as oxides and silicates such as sand, quartz, rock crystal, and mica. Pure silicon is obtained by heating silica and carbon in an electric furnace in the presence of carbon electrodes. Although there are many other methods, the Czochralski process is a common method to produce crystals of silicon used in solid-state semiconductors devices and micro-machined sensors.

1.4.2 Process of Developing Sensors

Sensors are developed using a wide range of materials and processing techniques. Methods for the fabrication of sensors vary depending on their specifications, size, and complexity. Many modern sensors are based on the properties of semiconductor materials and manufactured as microsensors or nanosensors.

Today, modern microsensors are fabricated by making full use of the properties of semiconductors and their associated technologies. In microsensors, the use of other materials and the deposition of thick and thin films are often required to give the sensing materials useful properties they would not normally have. For example, piezoelectric material films applied to a silicon wafer provide piezoelectric properties. There are several methods of depositing thin and thick films on substrates or semiconductor wafers, including spin-casting, vacuum deposition, sputtering, electroplating, and screen printing.

Spin-casting is often used for deposition of organic materials for humidity measurements and chemical applications. This process involves the use of a thin film material dissolved in a volatile liquid solvent. The solution is pored onto the sample and the sample is rotated at high speeds, thus spreading the material over the sample due to centrifugal force. As the solvent evaporates, a thin layer of the material remains on the sample. Although this method is adequate for most applications, in some cases the thin film may not be uniformly spread on the surface, which can cause nonlinearity and other problems.

In vacuum deposition, the material is converted into a gaseous form and deposited on the surface of the sample. A vacuum deposition system consists of a chamber and a heating mechanism, as heating of the material above the melting point is necessary. The gaseous atoms are allowed into the chamber in a controlled manner to be deposited on the sample. The thickness of the deposition is controlled by the evaporation time and the vapor pressure of the material. However, as in the case of spin-casting, distribution of the film on the surface may be nonuniform, thus causing problems associated with nonlinearity.

Sputtering is similar to vacuum deposition. In this method, an inert gas such as argon or helium is introduced into a chamber that contains anode and cathode electrodes supplied by an external high-voltage source. The anode contains the sample to be deposited on and the cathode contains the deposited material. The principle is that the high voltage ignites a plasma effect in the inert gas and the gas ions bombard the target containing the material to be deposited. When the kinetic energy of the bombarding ions is sufficiently high, some of the atoms from the target surface are freed and carried by the gas to the surface of the sample. The sputtering technique yields better uniformity, particularly in the presence of a magnetic field. This method does not require high temperatures, so virtually any type of material, including organic materials, or mixtures of different materials can be sputtered.

Chemical vapor deposition (CVD) is one of the most common methods used for the fabrication of semiconductor-based sensors. It is a widely applied technique, particularly in the production of optical and optoelectronic devices. The CVD process takes place in a reaction chamber where substrates or wafers are positioned on stationary or rotating tables. The dopants are allowed to enter the chamber mixed together with a carrier gas such as hydrogen. The substrate is kept at an elevated temperature that helps the additives to be deposited on the surface of the sample. The thickness of the deposition is controlled by the amount of dopant in the gas, the pressure at the inlet, and the temperature of the substrate.

1.4.3 Trends in Sensor Technology and IC Sensors

The present trend in sensor technology has shifted toward IC sensors in the form of microsystems, intelligent sensors, nanosensors, and others. Microsystems refer to the dimensions of devices in the micrometer (10^{-6} m) range, whereas nanotechnology refers to the dimensions of devices in the nanometer (10^{-9} m) range. Microsystems technology is well established and is simply known as MST. A subset of MST is microelectromechanical systems (MEMS). Another subset of MST is the microelectro-optical systems (MEOMS) and system-on-chip (SOC) devices. Most of the sensors manufactured by MEMS and MEOMS are three-dimensional devices with dimensions on the order of a few micrometers.

Data obtained from sensors and transducers are interpreted into forms that humans can understand by associated interface circuits. The very large scale integrated (VLSI) circuits have been extensively used to realize complex sensor modules (e.g., in the form of microsensors). Before the availability of microelectronics, sensors and transducers were coupled to external readout devices via suitable circuits. However, with the advent of microelectronics technology, sensors and transducers have developed in such a way that many processing components are integrated with the sensor on the same chip (broadly termed, IC sensors). Most IC sensors can interface to an exter-

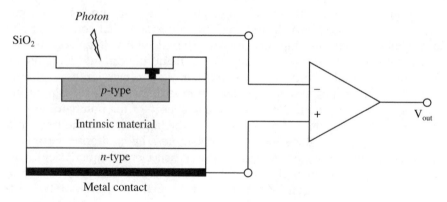

FIGURE 1.9
Typical structure of a photodiode.

nal microcontroller unit directly without any A/D conversion or other components. This is achieved either by inherent digital output sensors or by the integration of on-chip processing electronics within the sensing unit.

Semiconductor-based sensors are produced by using microfabrication techniques, which refers to the collection of processes used by the electronics industry for manufacturing ICs. IC sensors provide a simple interface, lower cost, and reliable input to electronic control systems. A typical example of an IC sensor is the photodiode, illustrated in Figure 1.9. In this sensor, the incident light falls on a reverse-biased pn junction and the photonic energy carried by the light creates an electron-hole pair on both sides of the junction, causing a current to flow in the circuit. The output voltage of photodiodes is highly nonlinear, thus requiring suitable linearization and amplification circuits, which can be included on the same chip.

Integrated circuit sensors can be grouped according to their signal domains:

- Radiant domain: sensors contain a wide spectrum of electromagnetic radiation, visible spectrum, and nuclear radiation. Some examples are photovoltaic, photoelectric, photoconductive, and photomagneto effect sensors.

- Mechanical domain: sensors include a wide range of devices from MEMS to tactile sensors. Some examples are piezoresistive, photoelectric and photovoltaic sensors, and micromachined devices.

- Thermal domain: sensors are largely semiconductor-based devices that exhibit sensitivity to temperature effects. Although sensitivity to temperature is undesirable in many applications, the temperature dependence of semiconductors can be useful for temperature measurements and control. Some of these devices are based on the Seeback and Nernst effects.

- Magnetic domain: sensors are made from magnetically sensitive semiconductors that are obtained by using doping techniques and thin films such as nickel-iron. The majority of sensors use principles of the Hall effect, magnetoresistance, and the Suhi effect.

- Chemical domain: sensors that include a large number of commercially available semiconductor sensors. These are based on techniques such as ion-sensitivity field effect transistors (ISFETs), chemically sensitive thin films, and polymers.

The usefulness of semiconductor-based IC sensors is enhanced considerably by the integration of microprocessors, microcontrollers, converters, logic circuits, and other digital components. Further, micromachining techniques combined with semiconductor processing technology provides a range of sensors integrated on the same chip for mechanical, optical, magnetic, chemical, biological, and other types of measurements. Advances in digital technology and cost-effective manufacturing techniques of IC sensors are expected to revolutionize instrumentation technology.

A typical example of an IC sensor is the microelectromechanical accelerometer, such as the ADXL150 and ADXL250 manufactured by Analog Devices. The ADXL150 is capable of sensing acceleration in a single axis, whereas the ADXL250 senses acceleration in two axes. These sensors include transducer elements and the necessary signal conditioning electronics together on a single IC. Both the ADXL150 and ADXL250 offer low noise (1 mg/Hz) and a good signal:noise ratio. The data obtained from each sensor can be acquired by suitable microcontrollers such as the PIC16F874, which has a 10-bit internal A/D converter. A transistor-based IC temperature sensor is illustrated in Figure 1.10.

Other examples of IC sensors are the power ICs (PICs). PICs are electronic devices that are already equipped with embedded internal sensors. They are produced by combining bipolar and metal oxide semiconductor (MOS) cir-

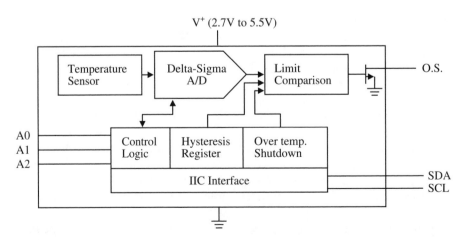

FIGURE 1.10
An IC temperature sensor.

cuitry with metal oxide semiconductor field effect transistor (MOSFET) technology. The approach to power ICs is based on the consolidation of a number of circuit elements into a single device. These devices would normally be discrete components, or a combination of standard and custom ICs with some discrete output device backups. In this single chip, some circuit elements (e.g., operational amplifiers, comparators, regulators) are best implemented by the bipolar IC technology. MOS circuitry handles logic, active filters, and time delays. Some circuits, such as A/D converters and power amplifiers, can be implemented by either bipolar or MOS technology.

An advantage of power ICs is that they are capable of directly interfacing between MCUs and system loads, such as solenoids, lamps, and motors. They provide increased functionality as well as sophisticated diagnostics and protection circuitry. Sensing of current levels and junction temperatures is a key aspect during normal operations for detecting several types of faults. Sensors within the PICs detect fault and threshold conditions, thus allowing the implementation of control strategies where various degrees of sensitivity are required for parameters such as temperature, current, and voltage.

Many other IC sensor systems consist of discrete or multiple sensors combined with application-specific ICs or some components of printed circuit boards. This leads to a diverse range of sensors requiring various forms of interface electronics. The interface requirements depend on the quantity to be measured, the types of physical effects, overall system architecture, and application specifications.

1.4.4 Sensor Arrays and Multisensor Systems

In many measurements, more than one sensor is required. Sensing arrays include a number of sensors for different measurands, such as pressure, flow, temperature, and vibration. These arrays are used to increase the measurement range, provide redundancy, or capture information at different times or different spatial points. A good example of a sensor array is in chemical applications where a single chip is used to measure different types of chemicals. Currently, considerable R&D effort is focused on multiple sensors or sensing arrays for the integration of all necessary signal conditioning components and computational capabilities on the same chip.

Complementary metal oxide semiconductor technology allows the integration of many sensors on a single chip, thus it is a common method applied in sensing arrays. Some examples of CMOS sensing arrays include photodiode arrays, ion detectors, moisture sensors, electrostatic discharge sensors, strain gauges, edge damage detectors, and corrosion detectors.

Photodiode arrays are a typical example of a sensor array. A photodiode consists of a thin surface region of p-type silicon formed on an n-type silicon substrate. A negative voltage applied to a surface electrode reverses the bias of the pn junction. This creates a depletion region in the n-type silicon, which contains only an immobile positive charge. Light penetrating into the deple-

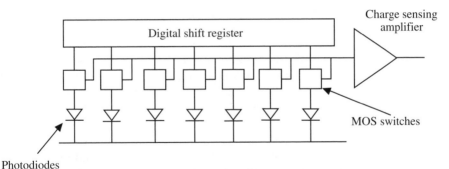

FIGURE 1.11
An IC image transducer.

tion region creates electron-hole pairs, which discharge the capacitor linearly in time. There are two basic types of photodiodes: serially switched photodiode arrays, shown in Figure 1.11, and charge-coupled photodiode arrays. In these arrays, the basic principle is to use light intensity to charge a capacitor and then read the capacitor voltage by shifting it through the registers. Such solid-state image sensors can be considerably complex when they are manufactured in IC forms.

Integrated multisensor chips are attracting considerable R&D attention. Multipurpose integrated sensor chips have been manufactured for the simultaneous measurement of physical and chemical variables. IC technology allows the design of complex systems on a single chip that incorporate high-performance analog subsystems such as op amps and data converters on the same die with digital circuits. These devices, generally manufactured by MOS technology, include signal conditioning, array accessing, and output buffering along with infrared sensing arrays, chemical sensors, accelerometers, vapor sensors, tactile sensing arrays, etc. Some of the multisensing functions of these chips utilize both the pyroelectric and piezoelectric effects of zinc oxide thin films.

Integrated microsensor chips constitute complex microsystems requiring very high performance microelectronic components together with nonelectronic miniaturized subsystems. These chips can be categorized as MOEMS, MEMS, lab-on-chip, radio frequency (RF) MEMS, system-on-chip, data storage MEMS, and so on. For each one of these it is possible to implement microsystems for many different functions, so the entire approach can be quite diversified.

1.4.5 Smart Sensors

In recent years, significant progress has been made in instruments and instrumentation systems because of the integration of microsensors, nanosensors, and smart sensors in measurement systems. A conventional sensor measures physical, biological, or chemical parameters and converts these parameters into electrical signals. These sensors require extensive external circuits and components for signal processing and display. The term smart sensor was adopted in the mid-1980s to differentiate a new class of sensors from conventional ones. Smart sensors have intelligence of some form and can convert a raw sensor signal into one that is much more convenient to use. It provides value added functions, thus increasing the quality of information rather than just passing the raw signal. Modern smart sensors can perform functions such as self-identification, self-testing, lookup tables, and calibration curves, and have the ability to communicate with other devices. All these additional functions are conducted by the integration of sensors with microcontrollers, microprocessors, or logic circuits on the same chip. The microprocessor contains RAM and ROM and can be programmed externally. Smart sensors also include signal amplification, conditioning, processing, and A/D conversion.

The integration of sensors with complex analog and digital signal processing circuits and microprocessors on the same chip has enabled extensive development of supporting software. The use of digital signal processing circuits and the integration of intelligent techniques such as artificial neural networks (ANNs) serve as nonlinear signal processing tools leading to convenient and easy to use devices. On-board operating systems and additional decision-making software such as artificial intelligence (AI) and complex logic circuits result in faster, more efficient, fault tolerant and reliable systems. A general scheme of a smart sensor is illustrated in Figure 1.12. In this example, the sensor is under microprocessor control.

A variety of smart sensors, also known as intelligent sensors, are manufactured with the neural network and other intelligence techniques programmed on the chip. These sensors are capable of assimilating large quantities of data and are capable of taking autonomous and appropriate actions to achieve goals in any dynamically changing environment. They are adaptable in anticipating events and complexities in the process, therefore sensing, learning, and self-configuration are key elements. Intelligent sensors

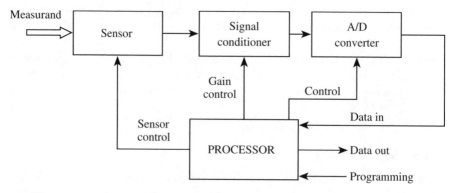

FIGURE 1.12
A block diagram of a smart sensor.

are available as pressure sensors and accelerometers, biosensors, chemical sensors, optical sensors, magnetic sensors, etc. Intelligent vision systems and parallel processor-based sensors are typical examples of such devices.

Artificial neural networks are used in many intelligent sensors, including single-sensor systems, redundant-sensor systems, multisensor systems, and fully integrated decision and control systems. In a single-sensor system, application of ANN results in improved system linearity beyond conventional IC compensation. In redundant-sensor systems, the accuracy and robustness measurements taken by identical sensors can be improved by ANN. In multiple-sensor systems, where different types of sensors are used for measuring different physical parameters, ANN improves the linearity of individual sensors and assists in complex decision making. The system configuration of a multisensor chip is shown in Figure 1.13. In fully integrated decision and control systems, ANN performs both sensor enhancement as well as intelligent control. Fully integrated systems find extensive applications in aerospace, defense, consumer products, and industrial needs.

Artificial neural networks have parallel architectures and can easily be simulated on digital computers. The network topologies, training algorithms, and optimize parameters are readily available in simulation packages. ANNs can also be implemented by hardware and software combinations using digital signal processor architectures, custom PC extension cards, array processors, and by application-specific IC design. Intel, Motorola, and others have released specialized ICs for general purpose neural network implementation. For cost-effective implementation of these ANN ICs, it is important to determine the existing technology that is most appropriate for the sensors in hand, the bandwidth requirements, and the training needs.

The NC3002 is an example of such a sensor, which is based on the digital VLSI parallel processing technique. These sensors are used in machinelearn-

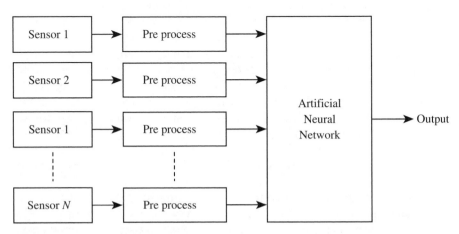

FIGURE 1.13
An intelligent multisensor chip configuration.

ing and image recognition supported by ANNs in quality determination and inspection applications. The architecture of the NC3002 is structured in a way that implements the Reactive Tabu Search learning algorithm, which is a competitive alternative to back propagation. This algorithm does not require derivatives of the transfer functions. The chip is suitable to act as a fast parallel number-crunching engine intended for operation with a standard CPU in single- or multiple-chip configurations.

Another example is the intelligent image sensors. These sensors are based on monolithic CMOS technology and contain on-chip A/D converters and appropriate microprocessor interface circuits. These types of sensors are an important part of digital cameras. They incorporate sensors, analog signal conditioning circuits, and memory elements on-chip. In digital camera applications, they are designed to operate in direct connection with a microprocessor bus.

1.5   Instrument and Sensor Communication and Networks

With recent advances in communication technology, instruments can easily be networked. Many processes require measurements of hundreds or perhaps thousands of parameters employing many instruments. The resulting arrangement for performing the overall measurement in a complex process is called the measurement system. In measurement systems, instruments operate autonomously, but in a coordinated manner. Information generated by each instrument may be communicated between the instruments themselves and controllers, or between instruments and other digital devices such as recorders, display units, printers, routers, base stations, or a host computer.

In complex measurement systems, digital instruments find wider applications for two main reasons: first, for their easy networking capabilities by means of remote communication methods such as RF, microwave, Internet, and optical techniques; and second, because of their on-board memory capabilities for data handling and storage. Transmission of data between digital devices is carried out relatively easily using wired or wireless transmission techniques. However, as the measurement system becomes large, communication can become very complex. To avoid this complexity, message interchange standards are used that are supported by appropriate communication hardware and software such as RS-232, universal serial bus (USB), EIA-485, and IEEE 488.

Today many instruments contain at least one RS-232 or USB port for communication purposes. Also, there are many companies offering RF RS-232 or USB systems for remote data transmission. It uses serial binary data interchange and applies specifically to the interconnection of data communication equipment (DCE) and data terminal equipment (DTE). DCE includes modems, which are devices that convert digital signals suitable for transmission through telephone lines. Relatively older technology RS-232 uses standard DB-25 connectors. With DS-25 connecters, although 25 pins are assigned, complete data transmission is possible by using only three pins—2, 3, and 7. The transmission speed can be set to a specific baud rate: 1200, 2400, 4800, 9600, 19,200, or 38,400 bits per second or higher. RS-232 can be used for synchronous or nonsynchronous communication purposes. The signal voltage levels are flexible, with any voltage between –3 V and –25 V representing logic 1 and any voltage between +3 V and +25 V representing logic 0. RS-232 was first issued by the Electronic Industries Association (EIA) and dates back to the 1960s. Since then it has evolved into a number of different types of pin configurations that need to be understood and modified for the system being used.

Many instruments also contain parallel ports because they are faster than their serial counterparts. Peripheral devices usually operate on parallel I/O ports. The architecture of a parallel bus defines the width of the data paths, transfer rates, protocols, cable lengths, and connector configurations. IEEE 488 is a common parallel bus that is used in variety of instruments and instrumentation systems and is suitable for monitoring and controlling clusters of instruments and other measurement systems. Another parallel bus, the small computer system interface (SCSI), connects high-speed computer peripherals, such as hard disk drives, to the main processor board.

In industrial applications, several standards for digital data transmission are available. These are commonly known as fieldbuses in the engineering literature. Some of these standards are widely accepted and used, such as the WordFIP, Profibus, Foundation Fieldbus, and LonWorks. The fieldbuses are supported by hardware and software (e.g., National Instruments chips and boards) that allow increases in data rates suitable with high-speed protocols.

1.5.1 Wireless Instrument Communication

Various techniques are used in wireless instrument communication, including optical and infrared methods, RF methods, and sonic methods. Since the main theme of this book is wireless instruments and networks, RF methods will be explained in detail.

Radio frequency system design is a multidisciplinary field that requires a good understanding and knowledge of many areas of disciplines including modern IC design and implementation. The understanding and knowledge areas necessary for the design and implementation of RF instruments and networks are illustrated in Figure 1.14.

In the design process, RF component architectures are planned according to available off-the-shelf components and ICs are selected to serve as many architectures as possible, leading to a great deal of redundancy at both the system and circuit levels. The components of a typical RF transmitter and base station are illustrated in Figure 1.15.

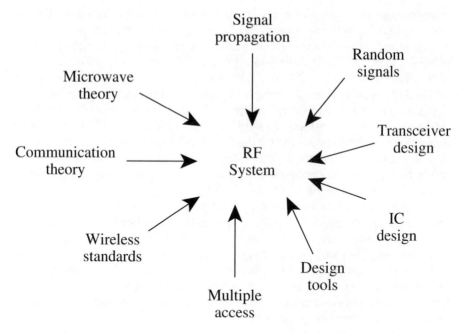

FIGURE 1.14
Disciplines required in RF design.

Analog and digital RF circuits are required to process signals that contain wide dynamic ranges at high frequencies. Signals have to be processed in suitable formats to make them ready for transmission. Issues such as noise, power, linearity, frequency, gain, and supply voltage need careful treatment and must be balanced against each other.

When data are transmitted over long distances, appropriate hardware and software must be used, such as modems, microwaves, or RF devices. On the receiving end, appropriate hardware and software interprets the received signals and extracts the information transmitted. Various modulation techniques are used to convert signals to suitable formats. For example, most modems, with medium-speed asynchronous data transmission, use frequency shift keyed (FSK) modulation. The digital interface with modems uses various protocols such as MIL-STD-188C to transmit signals in simplex, half-duplex, or full-duplex forms, depending on the direction of the data flow. The simplex interface transmits data in one direction, whereas full duplex transmits it in two directions simultaneously.

In the design process of RF circuits, the availability of specialized computer-aided analysis and synthesis tools may be limited. Therefore other methods must be used to simulate and model the circuits to observe the behavior of the complete system or parts of it.

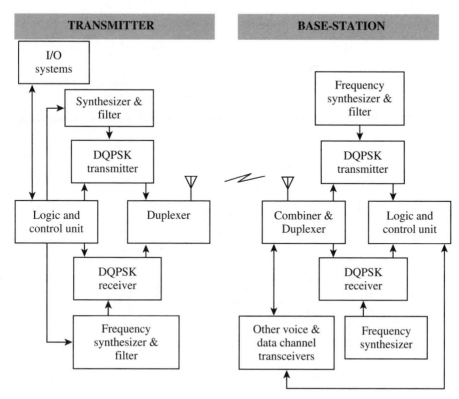

FIGURE 1.15
Components of a transmitter and base station in a wireless telephone system.

1.5.2 Modulation and Coding of Instrument Signals

For communication purposes, analog signals generated by instruments need to be converted to digital equivalents either by waveform coding, source coding, or combination of the two. Waveform coding involves the conversion of the amplitude of waveforms into binary equivalents to be modulated appropriately for transmission using a suitable modulation technique. Source coding simply models and samples selected features of a waveform. The receiver re-creates the original waveform from the received transmitted signals. Selection of a particular code for an application is a trade-off between available RF bandwidth and desired quality.

Waveform coding is done by sampling analog signals and converting them into digital form with an A/D converter. The operations of A/D converters are explained in Section 1.3. However, particularly in RF transmission, pulse code modulation can be used. Pulse code modulation translates a base band analog signal into a base band digital signal, as illustrated in Figure 1.16. The most commonly employed variations of pulse code modulation are pulse duration modulation (PDM), pulse position modulation (PPM), pulse

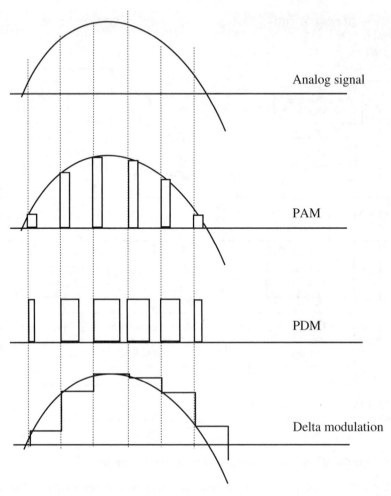

FIGURE 1.16
Examples of coding.

code modulation (PCM), pulse amplitude modulation (PAM), and delta modulation (DM).

In PDM, also known as pulse length modulation (PLM) or pulse width modulation (PWM), the amplitude information of the analog signal is represented by the duration of the sampled signal.

In PPM, also known as pulse phase modulation (PPM), uniform pulses are displaced in proportion to the amplitude of the original signal. This modulation method provides good synchronization of the transmitter and receiver. This method requires a DC shift to handle the negative amplitudes of the analog signal.

In PCM, analog signals are modulated in PAM and are then encoded into a binary code and transmitted as a digital stream. This method contains quantization errors and quantization noise, as explained in Section 1.3.

In PAM, the analog signal is sampled to obtain a pulse whose amplitude is proportional to the amplitude of the signal at the instant of sampling. This method is prone to noise and can be inefficient for transmission.

Delta modulation transmits changes in amplitudes of consecutive samples. It transmits a binary "1" for a current amplitude greater than the previous one and a "0" for an amplitude less than the previous one. For accuracy, the sampling step size (also called delta) is important for a true representation of the analog signal by the delta-modulated counterpart. If the step size is too small it leads to step overload, if it is long it leads to large quantization error, if it is too long it leads to granular noise.

Combinations of modulation techniques lead to adaptive (e.g., adaptive delta modulation (ADM)) and variable (e.g., variable slope delta modulation (VSDM)) techniques, resulting in greater efficiency and noise reduction.

Source coding compresses digital signals by using methods such as quantization, entropy coding, predictive coding, and transform coding. Quantization reduces the fixed-length PCM code words by using various compression techniques. Entropy coding assigns code words for frequently occurring messages, letters, and numbers. There are various forms, such as the Huffman code, arithmetic code, etc.

Predictive coding makes use of similar values in successive samples. It gives spatial and temporal prediction results by combining with quantization and entropy coding.

Transform coding converts blocks of samples into their frequency components, which can be represented by quantized and entropy codes for transmission. Most commonly used techniques involve wavelet transforms and discrete cosine transforms (DCTs).

1.5.3 Example of Wireless Communication Systems

Portable cellular phones are a good example of a well-established RF communication technology. Cellular systems are important in wireless instrumentation for two reasons: (1) there are operational instrumentation systems that use the cellular networks, and (2) the cellular technology directly flows to the instrumentation system. Therefore the knowledge and experience gained in cellular technology can be used in instrumentation systems.

Hundreds of cellular telephones can operate at the same time thanks to effective and efficient multiplexing techniques. There are three common types of multiplexing methods: frequency division multiple access (FDMA), time division multiple access (TDMA), and code division multiple access (CDMA). FDMA systems are largely analog and represent the first-generation devices. TDMA, representing the second-generation systems, was developed to increase the number of channels that can be provided in a limited bandwidth. CDMA systems use spread spectrum modulation techniques and provide further improvements in channel capacity. Table 1.1 illustrates the key characteristics of the main types of cellular systems currently in use.

TABLE 1.1

Characteristics of cellular telephone systems

Characteristics	AMPS North America	IS-54 North America	IS-95 North America	GSM Europe	PDC Japan
Frequency RX	869–894 MHz	869–894 MHz	869–894 MHz	935–960 MHz	940–956 MHz 1477–1501 MHz
Frequency TX	824–849 MHz	824–849 MHz	824–849 MHz	890–915 MHz	1429–1453 MHz
Access method	FDMA	TDMA/ FDMA	CDMA/ FDMA	TDMA/ FDMA	TDMA/FDMA
Duplex method	FDD	FDD	FDD	FDD	FDD
Channel	30	30	1250	200	25
Channel spacing (kHz)	832	832	10	124	1600
Bit rate (kbps)		48.6	1228.8	270.83	42
Modulation	FM	$\pi/4$ DQPSK	BPSK/ OQPSK	GMSK	$\pi/4$ DQPSK

In North America, the Federal Communications Commission (FCC) provide two groups of frequencies. Group A is used by nonwired common carriers, and group B is used by wired carriers. The channel width for both analog and digital cellular systems is 30 kHz. Thus a total of 832 transmit channels and 832 receive channels are provided in the 824–894 MHz band.

1.5.4 Examples of Wireless Sensors and Instruments

There are many companies offering a diverse range of wireless sensors and instruments. Companies have developed wireless solutions for their products and offer proprietary wireless sensors and instruments operating in licensed and unlicensed industrial, scientific, and medical (ISM) bands. Wireless systems are used for a wide variety of applications, including habitat monitoring systems for wildlife, environmental observation systems, and health monitoring systems. Chapter 4 and Chapter 5 provide many examples of wireless sensors, instruments, and their applications.

1.6 Industrial Instrumentation Systems

With the advances in communication technology, improvements in digital control techniques, and availability of modern sensors and transducers, there are many sophisticated industrial instrumentation systems that can be used in automation, manufacturing, process industries, and the like. In industrial operations, instrumentation networks and their control are broadly termed fieldbuses.

Fieldbuses are industrial communication systems that are configured and supported by single companies, or groups of companies, or private and government agencies. As an example of development of fieldbuses and standards, Fuji Electric of Japan has developed open systems in the field of instrumentation control and promoted the establishment of Foundation Fieldbus standards. The company has been working on field bus specifications, hardware development, promotions for increased usage, and public relations for many years. Fuji Electric has also started the Fiber-Optic Fieldbus Working Group, and now the results of the work have been finalized as a final specification in the Fieldbus Foundation. There many similar stories like Fuji Electric's in the history of common fieldbus systems.

However, in the course of fieldbus development, different technological approaches have been used, and there are more than 50 different fieldbus products or standards in the marketplace, which are reported by manufacturers in conferences, magazines, and exhibits. Table 1.2 lists some of the most commonly fieldbuses. Some of the important fieldbus products that are not mentioned in the list are the I/O-lightbus, IEC 61158, P-NET, INSTA

TABLE 1.2

Examples of commonly available fieldbuses

Research

MPS (Michigan Parallel Std., MSS)	University of Michigan
I²S	Delf University of Technology

Industrial

Foundation Fieldbus HSE/H1	Fieldbus Foundation
DeviceNet	Allen-Bradley
IEC/ISA SP50 Fieldbus	Fieldbus Foundation
LonTalk/Lonworks	Echeleon Corp.
Profibus DP/PA	Siemens
InterBus-S	InterBus-S club, Phoenix, AZ
WorldFIP	WorldFIP

Automotive

CAN	Bosch
MI-Bus	Motorola
J-1850, (CAN),J2058, J2106	SAE, Chrysler, General Motors

Home automation

Smart House	Smart House LP
CEBus	EIA

Building and Office Automation

BACnet	Building Automation Industry
IBIbus	Intelligent Building Institute

(EIB), Modbus, Bitbus, Arcnet, and Highway Addressable Remote Transducer (HART).

Fieldbuses enable the interconnection of field devices among themselves or to computers or other digital devices. Fieldbuses thus represent industrial systems integrating analog and digital field devices for information exchange and control purposes. In doing so, fieldbuses bridge the gap between the two by means hardware and software.

The fieldbus protocols are largely developed for distributed processes with a centralized control philosophy (as in the IEC/ISA fieldbus and FIB) or distributed processes with a distributed control philosophy (as in Profibus). At the moment, there are many fieldbuses offered by a wide range of vendors driven by commercial interests or application-specific requirements. However, there is a worldwide effort to standardize fieldbus protocols so that they become independent of vendors and application-specific requirements. Although this was realized by the IEC 61158 standards in 2000, its effect and wide acceptability is yet to be seen. A common fieldbus standard will allow the implementation of universal control strategies to improve intrinsic safety; provide better security; enable easy installation, maintenance, monitoring, and diagnostics of the equipment; provide interoperability; and yield cost reductions for vendors and ultimately cost savings for end users. Emerging wireless technology is likely to replace most of the existing fieldbuses.

The IEC 61158 industrial fieldbus standard is based on at least six standards: ControlNet, Profibus-PNet, Foundation Fieldbus, Swiftnet, WorldFIP, and Interbus. The standard is subdivided into seven parts. Part 1 and Part 2 are for general introduction and management issues. Part 3 through Part 6 cover the data link to application layers in the International Standards Organization/Open System Intercommunication (ISO/OSI) model. IEC 61158 is designed mainly for ensuring reliable interdevice communication. However, standard function blocks and enhancement of programming language support at the user layer are covered by other standards such as IEC 61131-3, IEC 6184, and IEC 61499.

1.6.1 Industrial Communication Networks

The fieldbuses interconnect field devices, as shown in Figure 1.17. In a complete fieldbus configuration, a variety of components such as sensors, instruments, analog and digital components, and controllers are connected to the system for measuring, monitoring, and controlling the process. These interconnected field components can be grouped into three major categories: (1) traditional analog and discrete I/O devices, (2) hybrid analog and digital devices, and (3) purely digital devices. The analog devices are usually connected to controllers by means of dedicated 4 mA to 20 mA analog current loops. The hybrid components are capable of using both analog and digital communication techniques. For example, a protocol called HART uses digital communication signals on conventional 4 mA to 20 mA analog signals. The

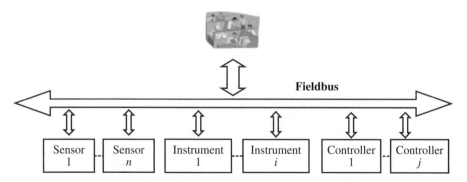

FIGURE 1.17
Connection of sensors, instruments, and controllers to a fieldbus.

third type, purely digital components, usually requires digital interface hardware and software such as RS-232, USB, IEEE 488 parallel buses, and so on.

In the wake of relatively new technology, part of the communication may be realized by new wired techniques such as I²C or I²S (Integrated Smart Sensor) buses, and many others. These buses have the following features:

- They can be simple, requiring only two communication lines (clock line and data).
- They require a minimum of electronic components for full operation.
- Transmission can be terminated either by master or slave.

Today many instruments and intelligent sensors communicate using wireless techniques with protocols and standards such as Bluetooth, ultra-wideband (UWB) radio frequency, 802.11a/b/g, general packet radio service (GPRS), ZigBee, 802.15.4, and IEEE 1451.5. RF communication can be realized by standardized or nonstandard methods. The choice of these methods depends on the application requirements, the complexity of the instruments and sensors, and the capabilities of the host system.

1.6.2 Basic Elements of Industrial Sensor Networking

Today, because of the cost and availability of advanced microprocessors and their supporting components, many sensors and instruments are based on powerful digital systems integrated at various stages from sensing elements to complex communication networks. Therefore microprocessors, microcontrollers, and digital signal processors are essential elements for interconnection to the fieldbuses. The use of microprocessors allows the sensors and controllers to be self-supporting and semiautonomous by enabling them to make local decisions under the instructions of the central controller. Microprocessors also provide a convenient man-machine interface, diagnostic capabilities, self-configuration, and self-identification.

Almost all modern fieldbuses operate on digital and computerized systems. A majority of sensors still produce analog signals, thus requiring A/D conversion. Similarly many controllers require analog signals, thus digital signals from the bus need to be converted back into analog forms. Typical elements needed for the connection of sensors to a digital environment are illustrated in Figure 1.18 and explained in Section 1.3.4. The front-end analog signal processing is necessary for amplification or attenuation and filtering of signals generated by the sensors. In addition to analog signal processing, the A/D conversion process takes place with the aid of appropriate S/H devices and multiplexers. Multiplexers allow the connection of many sensors and transducers to the same signal-processing media. For the process actuators and controllers, the digital signals are converted to analog waveforms by D/A converters and are then filtered and amplified before being applied to the controllers.

1.6.3   Industrial Network Protocols

Field instruments are connected to digital systems and computers, thus forming networks such as direct digital control (DDC) systems, supervisory control (SC) systems, distributed control systems (DCS), hybrid control systems, and supervisory control and data acquisition (SCADA) systems. The interconnected devices communicate through a particular common bus network using communication protocols that are developed to guarantee precise intended information flow between components. In industrial systems, different control units may need to share the data collected by a sensor and disseminate the processed data to various actuation units. An example is the shutdown of many units in a nuclear plant if a hazard is sensed. In this respect, effective communication of devices operating in groups has particular importance. These types of communication are realized generally by the fast data link layer protocols operating within fieldbus systems. These protocols provide a series of facilities, which help the application process to check that the real-time operation requirements are met.

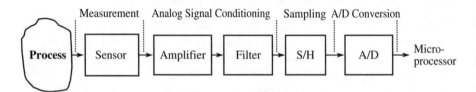

FIGURE 1.18
The basic elements of sensors.

During the communication process, the messages that sensors generate can be divided into short messages and long messages. While short messages are usually processed locally, long messages are processed by central computers or other remote devices. The length and type of message are fairly important in the configuration of fieldbuses. The fieldbuses are the first stage of information gathering, data flow, and decision making about the process under control. Some examples are the integration of the fieldbus with a computer-aided facility management (CAFM) system and integration of fieldbus systems into distributed object-oriented systems.

The information flow between nodes, individual instruments, and computers is regulated by protocols. According to the Institute of Electrical and Electronic Engineers (IEEE), a networking protocol is "a set of conventions or rules that must be adhered to by both communicating parties to ensure that information being exchanged between two parties is received and interpreted correctly." Most industrial communication systems use the ISO/OSI seven-layer reference model, simply known as the OSI reference model or OSI stack. Details of the OSI model are provided in Section 3.3 and Section 3.4. The OSI reference model has seven layers, of which each is an independent functional unit. Each layer uses functions from the layer below it and adds its own requirements to provide functions for the layer above. The lowest three layers are network-dependent layers, the highest three layers are network-independent layers (application oriented), while the middle layer (transport layer) is the interface between the two. Most fieldbuses use only a few of these layers, these being the physical, data link, network, and application layers.

1.6.4 The Ethernet and Fieldbuses

A significant technological shift in industrial automation in recent years has been the adoption of fieldbuses and the emergence of PCs on the plant floor. This strategy has lead to the wide dissemination of information over Ethernet industrial networks via transmission control protocol/Internet protocol (TCP/IP) ports. A typical modern industrial system with many cells connected together over an Ethernet is illustrated in Figure 1.19. Each cell consists of one or more fieldbuses, PCs, programmable logic controllers (PLCs), and other necessary elements of a DCS.

TCP/IP over an Ethernet is heralded as the industrial communication protocol of the future. This is because Ethernet provides many benefits such as lower costs and deployment of more devices in a given network. It is well understood and widely applied. An Ethernet increases the interoperability of a diverse range of equipment and allows greater integration of business systems with the plant floor. Expanding the enterprise's office communication network has well-known cost and efficiency benefits. Real-time access to manufacturing data can increase throughput and reduce inventories,

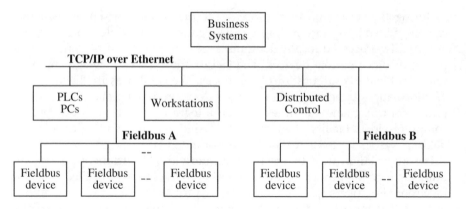

FIGURE 1.19
Typical modern industrial instrumentation systems.

while tracking the progress of customer orders can provide visibility to clear bottlenecks, as well as aid in safety and maintenance.

The implementation of fieldbuses with an Ethernet and the Internet leads to remote control of plants and the entire systems. There are many operational examples of systems that combine the benefits of the two to create a synergy. It appears that some of the protocol originally designed for the Internet can be used for this purpose. This combination provides access to the fieldbus system from local or global computer networks anywhere in the world.

1.6.5 Implementation of Fieldbuses

The implementation of fieldbuses involves hardware as well as software, with a trade-off between the two. Most fieldbuses have associated chips for implementing all or portions of system parameters. For example, in the LonWorks protocol, all layers can be implemented by chips and associated software. The software defines all the objects and libraries that are readily available for the customer. But for CAN, the lower two layers are implemented by a protocol chip (82C200), while all the software has to be implemented by the user.

Implementations of fieldbuses vary from system to system as vendors offer different forms of supporting hardware and software. Also, the nature and requirements of processes differ from one another. In order to explain the typical characteristics of fieldbus implementation, Foundation Fieldbus will be used as an example.

Foundation Fieldbus is an all-digital, serial, two-way communication system that interconnects field equipment such as sensors, actuators, controllers, and computers. It is essentially a local area network (LAN) for instruments used in manufacturing, automation, and process industries with built-in capability to distribute the control application across the network. It is supported by an independent nonprofit corporation known as the Fieldbus Foundation, consisting of more than 100 of the world's leading suppliers, manufacturers, and end users. Foundation Fieldbus, created in 1994, is a joint effort of two major organizations; the Interoperable Systems Project (ISP) and WorldFIP/North America. Its committee is formed by the members of the International Electrotechnical Commission (IEC) and Industry Standard Architecture (ISA) backed up by Profibus of Germany and FIP of France.

Foundation Fieldbus is particularly suitable in systems where the traffic involved is the periodic exchange of short information frames. Nevertheless, purely periodic exchange of short information frames is not sufficient to meet all the communication requirements of the process. In this respect, the implementation of the communication models offered by the application layer through the access medium mechanisms are made available at the data link level. Therefore asynchronous message (e.g., alarms, operator commands) transmissions are allowed as well as synchronous transmission. This approach is the key point in the definition of this particular protocol, as only correct mapping between application layer communication models and data link layer mechanisms can guarantee the time-critical requirements of these processes.

An important characteristic of the Foundation Fieldbus is the function blocks that define the inputs, outputs, mode structures, and event subsystems for common control systems. These blocks provide a consistent and easy-to-use method for modeling standard control system functions. As far as the control computers are concerned, the system is seen as a collection of function blocks rather than a collection of instruments. An important contribution of the Foundation Fieldbus is the flexibility of creating field devices that are capable of storing and executing software routines in their own way. With the use of such intelligent devices, the function blocks permit the distribution and integrated execution of process functions among the field devices themselves. This leads to advanced capabilities in this fieldbus that provides plant asset management (PAM) for maintenance and efficient operation of the entire plant.

The Foundation Fieldbus topology comprises two basic parts: a lower speed fieldbus called H1 and a higher speed fieldbus called H2. These two fieldbuses can be connected together via a device called a bridge. The device descriptions are a standard mechanism that allows a control system to acquire the definitions of messages from a field device. Foundation Fieldbus is equipped with information on the device characteristics, such as manufacturer's details, supporting software capabilities, available function blocks, and diagnostic capabilities. Such information is developed in device description language (DDL), which originated from the HART protocol. This approach is expected to enable vendors to describe the functions of their devices in a standard way. The manufacturers can also add new capabilities and novel features to their devices that can be easily accessible by already installed systems and new users.

1.6.6 Design and Application Examples of Fieldbuses

Fieldbuses are used in a wide variety of industries from cars to complex manufacturing plants consisting of tens of thousands of sensors, transducers, and controllers. Many companies offer operable products only with a particular fieldbus or a few selected ones. However, there is a considerable push to produce interoperable products that can be used with a variety of fieldbuses. As an example, the Systems Integration Specialists Company (SISCO) Sterling Heights, MI, offers a broad range of fieldbus products including adapter cards, servers, interface boards, programming tools, bus analyzers, protocol source codes, and so on, that are suitable for a wide range of fieldbuses. This company also offers measuring devices such as pressure transmitters, temperature transmitters, programmable logic controllers, fieldbus-to-current converters, fieldbus-to-pneumatic signal converters, valves, and other accessories. They also have products for wireless connections for accessing distant sensors and actuators.

There are many manufacturers offering various ICs and chip sets for fieldbus systems at competitive prices. Local operating network (LON) chips, for example, are available from Motorola and Toshiba for less than $5. However, based on speed and response, CAN chips are reliable and are used extensively in the automobile industry. They are manufactured by Intel, Motorola, Philips, and others.

In a laboratory environment, the author implemented a CAN protocol, as illustrated in Figure 1.20. CAN is a two-layer protocol covering the physical and data link layers of the ISO/OSI model. To implement a CAN-based network, an application layer must be defined and network connections must be specified. This particular system in the laboratory contains a number of sensors and actuators with an emphasis to safety-related issues. Among many other CAN controllers, the 82C200 chip from Philips was selected.

In this project, the implementation was realized by three sublayers: the logical link control (LLC) and the medium access control (MAC) sublayers for the data link layer. This CAN implementation allowed messaging between sensors, actuators, man-machine interfaces, controllers, and other control activities. This architecture provided a communication environment upon which a high-speed, real-time centralized or distributed control platform could be created.

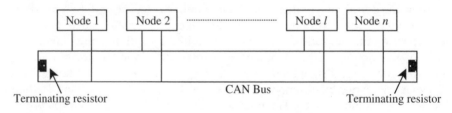

FIGURE 1.20
A typical CAN network layout in a laboratory environment.

Designing a fieldbus system requires careful understanding of the characteristics of the process to be monitored and controlled. Once the measuring, monitoring, and control objectives are identified, selection of the appropriate fieldbus will depend on the costs of the system and the ease of integration of the selected fieldbus to the existing instruments in the plant. As indicated in Table 1.2, suitable fieldbuses must be selected for a particular application, such as BACnet for intelligent buildings, Profibus for industrial applications, and so on. Once the decision is made, manufacturers' instructions for installation, commissioning, and maintenance must be followed closely. It is important that in the selection process the interoperability of sensors and instruments offered by different manufacturers be considered.

Implementation of fieldbuses may involve many choices, depending on the level of applications. Figure 1.21 shows the application scope of several buses in industrial automation and process control markets. The fieldbuses can be implemented using conventional wired methods or using new wireless techniques. Wireless implementations of fieldbuses are becoming popular as many new instruments have wireless communication features. The most popular means of transmission using wireless systems are LAN adapters, radio modems, and packet controllers. The difference between these is in the way the data are processed between and during transmission.

Wireless transmission devices are implemented to create complete wireless networks or partial networks with wireless segments in wired networks. Simple fieldbuses such as those based on RS-232C standards (e.g., Modbus) can easily be realized with the use of radio modems or packet controllers.

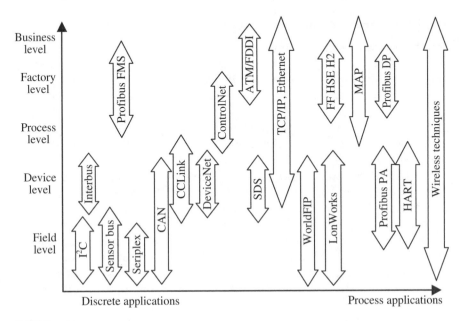

FIGURE 1.21
Application scopes of several fieldbuses used in industrial environments.

If the network is more complicated or the required transmission speed is higher, LAN cards and more powerful processors are used. A number of reports have been published (e.g., Liptak: 2002) describing methods for wireless transmission in networks such as LonWorks, CAN, and FIP. Wireless transmission is also included in the Foundation Fieldbus standard.

Another fieldbus implementation uses fuzzy component networks such as fuzzy sensors, fuzzy actuators, and fuzzy inference on fieldbuses. These fuzzy components are formed in groups to form fuzzy cells. Depending on its configuration, each fuzzy cell is able to perform the functions of one or several fuzzy components. Fuzzy cells are characterized by their ability to handle and exchange fuzzy symbolic information. Their configuration is performed via the fieldbus using languages such as the Prototype of Integrated Language for Symbolic Actuators and Sensors (PLICAS), which is specially designed for this purpose. A PLICAS compiler is integrated inside each cell, thus conferring interoperability properties to the cell.

In time-critical communication systems, the Hopfield neural network (HNN) is extensively used. In these systems, the correct implementation of the proper scheduling sequence is very important. Generally the length of the scheduling sequence is equal to the minimum common multiple of the periods of the processes, and needs to be memorized for later implementation. During this process, some problems arise if the sequence is large. The computational complexities arising in the sequences are overcome by adopting a HNN integrated with the fieldbus.

1.6.7   Telemetry and SCADA Systems

Telemetry is a wireless communication technique for gathering information at some remote location and transmitting the data to a convenient location. Telemetry can be accomplished by different methods: optical, mechanical, hydraulic, electrical, etc. More recently, the use of optical fiber systems has allowed the measurement of broad bandwidth, with high immunity to noise and interference. Other proposed telemetry systems are based on ultrasound, capacitive or magnetic coupling, and infrared radiation, although these methods are not routinely used.

The discussion in this section will be limited to the most common telemetry systems, which are based on electrical signals and wireless transmission of information. In many industrial applications, wireless telemetry is more complex as it requires RF or microwave energy for transmission and reception of information. Wireless transmission is used in those applications where measurement areas are not normally accessible or the system transmitting or receiving the information is mobile. Some wireless telemetry units have enough capacity to transmit multiple channels of information simultaneously.

Figure 1.22 shows a generic telemetry system. The main parts and their functions in a telemetry system are

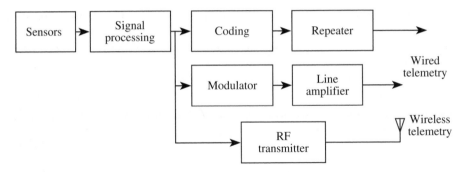

FIGURE 1.22
Basic components of a telemetry system.

- Transducers to convert physical variables to be measured into electrical signals that can be easily processed.
- Signal conditioning circuits to amplify the low-level signal from the transducer, limit its bandwidth, and adapt impedance levels.
- Subcarrier oscillators whose signals are modulated by the output of the different transducers once they are processed and adapted.
- Coding circuits, which can be a digital encoder, an analog modulator, or a digital modulator, that adapt the signal to the characteristics of the transmission channel (i.e., a wire or an antenna).
- Radio transmitter, in wireless telemetry, modulated by the composite signal.
- Impedance line adaptor, in case of wire transmission, to adapt the characteristic impedance of the line to the output impedance of the circuits connected to the adaptor.
- Transmitting antenna.

The receiving end consists of similar modules. For wireless telemetry, these modules are

- Receiving antenna designed for maximum efficiency in the RF band used.
- Radio receiver with a demodulation compatible with the modulation scheme.
- Demodulation circuits for each of the transmitted channels.

The transmission in telemetry systems, and wireless ones in particular, is done by sending a signal whose analog variations in amplitude, frequency, or phase are known functions of the signals from the sensors or transducers. Digital telemetry systems send data digitally as a finite set of symbols, each one representing one of the possible finite values of the composite signal at the time it was sampled.

Effective communication distance in a wireless system is limited by the power radiated in the transmitting antenna, the sensitivity of the receiver, and the bandwidth of the RF signal. As the bandwidth increases, the contribution of noise to the total signal also increases, and consequently more transmitted power is needed to maintain the same signal-to-noise ratio. This is one of the principal limitations of wireless telemetry systems. In some applications, the transmission to the receiver is carried out by the base band, which is located after the conditioning circuits. The advantage of base band telemetry systems is their simplicity. Base band transmission is normally limited to only one channel.

Almost all digital instruments are good candidates for integration in an instrumentation network by a telemetry link. Telemetry is widely used in space applications for measurements of distant variables and control of actuators. In most of these applications, for example, in space telemetry, it is very important to design systems with minimal power consumption. Some vehicles and trains in transport systems use wireless telemetry for monitoring and control. In clinical practice, the use of telemetry for patients increases their quality of life and their mobility, as the patients no longer need fixed monitoring systems. Several medical applications are based on implanting a sensor in a patient and transmitting the data for further analysis and decision making.

Multichannel telemetry is used to measure many different processes and physical variables in various applications. In these multiple measurements, base band telemetry is not an option, as it would require building a different system for each channel. Multichannel telemetry is achieved by sharing a common resource (transmission channel), as is shown in Figure 1.23. Sharing of the transmission channel by all the measurement channels is achieved by multiplexing.

Telemetry uses mainly two types of basic multiplexing techniques: FDMA and TDMA. In FDMA, different subcarrier frequencies are modulated by the different measurement channel signals, which shifts the information spectrum from the base band to the subcarrier frequency. Then the subcarrier

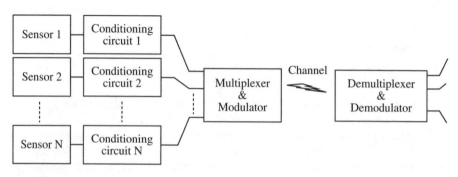

FIGURE 1.23
Block diagram of a multichannel telemetry system.

frequency modulates the RF carrier signal, which allows the transmission of all desired measurement channels simultaneously. In TDMA, the whole channel is assigned to each measurement channel, although those channels may be active only a fraction of the time. TDMA techniques use digital modulation to sample the different measurement channels at different times.

The term SCADA has been adopted by the process control industry to describe a collection of computers, sensors, and other equipment interfaced by telemetry technology in order to monitor and control processes. The uses of SCADA systems are almost endless, as they are only limited by the designer's imagination. These systems are used to monitor the status of high-voltage power distribution lines and stations, RF energy delivered by broadcasting stations, water levels in reservoirs, and so on. As digital systems become more sophisticated, the use of remote telemetry and control systems will become essential for adequate management of resources.

One of the advantages of using SCADA systems is that the existing sensors in current applications can easily be incorporated into the overall system, thus decreasing the economic cost of migrating to a new technology. In doing so, a SCADA system adds a new level of features and provides additional capabilities to current systems. Furthermore, system control can be moved from various remote locations to a centralized location.

1.7 Noise and Distortion

Noise is a serious problem in communication systems. Noise can be defined as random and unpredictable internal and external electrical signals produced by natural processes and interference. Internal noise can be caused by thermal energy, noise generated within devices, and so on. External noise is caused by atmospheric effects such as thunderstorms and solar activity. Interference is noise produced by man-made sources such as transmitters from other communication systems, power lines, and machinery.

1.7.1 Internal Noise in Electronic Systems

Noise is an important concept in all types of electronic devices. Noise is unwanted signals generated internally by the device itself or by signals that are externally imposed upon it. Noise can be inherent within the circuit or it can be interference picked up from outside the circuit. There are many different types of inherent and interference noise. Depending on the device, some noise may be significant, whereas other noise may not be significant at all. Since noise is a concern for all devices, a general discussion of noise will be given. There are many different types of inherent noise, such as thermal noise, shot noise, excess noise, burst noise, partition noise, recombination noise, and spot noise.

Thermal noise, also known as Johnson noise, is generated by the random collision of charge carriers with a lattice under conditions of thermal equilibrium. Thermal noise can be modeled by a series voltage source, v_t, or a parallel current source, i_t, in terms of mean-square values:

$$\overline{v_t^2} = 4kTR\Delta f \qquad \text{and} \qquad \overline{i_t^2} = \frac{4kT\Delta f}{R} , \qquad (1.7)$$

where T is the temperature in Kelvin, R is the resistance in ohms, and Δf is the bandwidth in hertz over which the noise is measured. The equation for $\overline{v_t^2}$ is known as the Nyquist formula. The amplitude distribution of the thermal noise can be modeled by a Gaussian probability density function.

Shot noise is caused by the random emission of electrons and by random passage of charge carriers across potential barriers. The shot noise generated in a device is modeled by a parallel noise current source. The mean square shot noise current in the frequency band Δf is given by

$$\overline{i_{sh}^2} = 2qI\Delta f , \qquad (1.8)$$

where I is the DC current flowing through the device and q is the electron charge. This equation is commonly referred to as the Schottky formula. Like thermal noise, shot noise is white noise with a uniform and Gaussian distribution.

Excess noise is caused by the variable contacts between particles of the resistive material. It is known that carbon-composition resistors generate the most excess noise, while metal-film resistors generate the least. The mean square excess noise current can be written as

$$\overline{i_{ex}^2} = \frac{10^{N_I/10}}{10^{12} \ln 10} \times \frac{I^2 \Delta f}{f} , \qquad (1.9)$$

where I is the DC current through the resistor and N_I is the noise index.

Burst noise, also known as popcorn noise, is caused by a metallic impurity in pn junctions. When amplified and reproduced by a loudspeaker, it sounds like corn popping. On an oscilloscope, it appears as fixed-amplitude pulses of random varying width and repetition rate. The rate can vary from one pulse per second to several hundred pulses per second.

Partition noise occurs when the charge carriers in a current have a possibility of dividing between two or more paths. The noise is generated in the resulting components of the current by the statistical process of partition.

Partition noise occurs in bipolar junction transistors (BJTs), where the current flowing from the emitter into the base can take one or two paths.

Generation-recombination noise is generated in semiconductors by the random fluctuation or free carrier densities caused by spontaneous fluctuations in the generation, recombination, and trapping rates. This noise appears both in BJTs and field effect transistors (FETs), but not in MOSFETs.

Spot noise is the root mean square in a band divided by the square root of the noise bandwidth. For a noise voltage and current it has the units of $V\sqrt{Hz}$ and $A\sqrt{Hz}$, respectively. For white noise, the spot noise in any band is equal to the square root of the spectral density. The spot noise voltage of a device output is given by $\sqrt{v_{no}^2 / B_n}$. Often the noise bandwidth is expressed in the context of a second-order band-pass filter, given by $B_n = \pi B_3 / 2$, where B_3 is the –3 dB bandwidth. In the case of band-pass filters with two poles at frequencies f_1 and f_2, the noise bandwidth is given by $B_n = \pi(f_1 + f_2)/2$.

Total noise is a combination of the noise current and noise voltage. In total noise, the instantaneous voltage ($v = v_n + i_n R$, where v_n is the noise voltage and i_n is the noise current) plays an important role. The mean square voltage of the total noise can be expressed as

$$\overline{v^2} = \overline{\left(v_n + i_n R\right)^2} = \overline{v_n^2} + 2\rho\sqrt{\overline{v_n^2}}\sqrt{\overline{i_n^2}}R + \overline{i_n^2}R^2 \,, \tag{1.10}$$

where ρ is the correlation coefficient defined by

$$\rho = \frac{\overline{v_n i_n}}{\sqrt{\overline{v_n^2}}\sqrt{\overline{i_n^2}}} \,. \tag{1.11}$$

For the case $\rho = 0$, the sources that generate the noise are said to be uncorrelated or independent.

In many circuit analyses, noise signals are represented by phasors in the form of complex impedances. Also, the noise generated by electronic devices, such as amplifiers, can be modeled by referring all internal noise sources to the input.

Noise bandwidth is the bandwidth of a device having a constant passband gain, which passes the same root mean square noise voltage when the input signal is white noise. White noise has the same amplitude for a given frequency range. The noise bandwidth is given by

$$B_n = \frac{1}{A_{vo}^2} \int_0^\infty \left| A_v\left(f\right) \right|^2 df \,, \tag{1.12}$$

where A_{vo} is the maximum value $|A_v(f)|$. For a white noise input voltage with a spectral density $S_v(f)$, the mean square noise voltage of the filter output is $\overline{v_{no}^2} = A_{vo}^2 S_v(f) B_n$.

The noise bandwidth can be measured with a white noise source with a known voltage spectral density. If the spectral density of the source is not known, the noise bandwidth can be determined by comparing it with another source with a known noise bandwidth. The noise bandwidth of the unknown source can be expressed by

$$B_{n2} = B_{n1} \frac{\overline{v_{o2}^2}}{\overline{v_{o1}^2}} \left(\frac{A_{vo1}}{A_{vo2}} \right)^2. \tag{1.13}$$

The white noise source must have low output impedance so that the loading effect does not change the spectral density of the source.

The spectral density of a noise signal is defined as the mean square value of noise per unit of bandwidth. For example, for thermal noise generated by a resistor, the voltage spectral density can be expressed as

$$S_v(f) = 4kTR. \tag{1.14}$$

As can be seen from Equation 1.14, since the spectral density is independent of the frequency, the thermal noise is known to have a flat or uniform distribution, and therefore is called white noise.

1.7.2 Interference

Noise can also be transmitted to electronic circuits by external sources. There are several classifications of transmitted noise, depending on how it affects the output signal and how it enters the electronic circuit. The transmitted noise can be magnetic, capacitive, or electromagnetic. The interfering noise can be additive or multiplicative with respect to the output signal.

In additive noise, the noise magnitude is independent of the magnitude of the useful signal. Clearly, if there is no signal from the device, only the noise is observed. Additive noise is expressed as

$$v_{\text{out}} = v_{\text{signal}} + v_{\text{noise}}. \tag{1.15}$$

In multiplicative noise, the signal becomes modulated by the transmitted noise. It grows together with the magnitude of the signal. Multiplicative noise is expressed as

$$v_{\text{out}} = v_{\text{signal}} + v_{\text{signal}} \times v_{\text{noise}}. \tag{1.16}$$

Transmitted noise can be periodic or random, depending on where and how it is generated. It can be inductive due to magnetic pickup or capacitive due to electrostatic pickup or electric charges, and at high frequencies transmitted noise can be electromagnetic, or various combinations of all three.

The offending external sources can be identified from the noise picked up by the instrument. For example, 50/60 Hz power supplies typically generate about 100 pA noise signals, whereas 120/100 Hz power supplies can generate few microvolts. Radio broadcast stations can induce noise voltages up to 1.0 mV. Mechanical vibrations can generate currents from 10 pA to 100 pA.

1.7.3 Noise and Distortion in Communication Systems

Noise and distortion are common problems in all types of communication systems. Noise in communication systems comes from factors such as thermal agitation, RF interference, cross talk, and erratic switch contacts. In communication systems, all extraneous signals appearing at the channel output that are not due to input signals are considered noise.

In many RF applications, noise is considered to be Gaussian, even though in practice this may only be an approximation. Noise is said to be Gaussian if it is random and is the result of many components of noise overlapping with each other. The probability density function of Gaussian noise is the familiar bell curve.

In wireless communication systems, the received signal must be substantially greater than the RF noise. There are many noise sources (Figure 1.24), including internal noise (generated by the receiver and supporting electronic components) and external noise (man-made and natural). It is known that at lower frequencies, less than 30 MHz, external noise received by the antenna from either man-made or natural noise is greater than internal noise. At frequencies greater than 30 MHz, internally generated noise is generally greater than external noise received by the antenna. Without filtering, the noise will have a wide range of frequency components.

Noise power is expressed by the equation

$$P_n = kTB, \tag{1.17}$$

where P_n is the available noise power in watts, T is the noise temperature in Kelvin, k is the Boltzmann constant (1.38×10^{23}), and B is the effective receiver noise bandwidth in Hertz.

Man-made noise is mainly due to electric motors, neon and fluorescent lights, power lines, ignition systems, domestic appliances, industrial equipment and appliances, and digital systems operating in the vicinity. In addition, there may be much noise due to other communication systems. The interference may include signals on assigned frequencies, and harmonics and subharmonics associated with those frequencies. Generally man-made

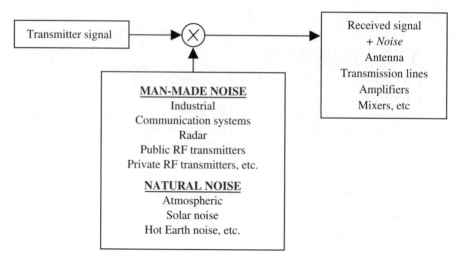

FIGURE 1.24
Natural and man-made noise sources in communication systems.

noise is assumed to decrease with increasing frequencies and can be expressed as

$$T_a = 100 \times T_a/f_{MHz},\qquad(1.18)$$

where T_{100} is the man-made noise temperature at a frequency of 100 MHz.

To determine man-made noise at a given site, it is necessary to make noise measurements using specialized equipment. The measurements may differ by time of day, season of the year, direction of measurements, and intensity of noise generated in the vicinity at the time of measurements.

The required signal:noise ratio for a communication system depends on the type of modulation used and the application specifications. In analog communication systems, the noise level that can be tolerated can be higher, thus giving a high signal:noise ratio. Digital systems may not tolerate high noise levels, but information from the received signals may be extracted better due to application of digital techniques. In any case, the signal:noise ratio can be improved by the use of quality components, careful design, and applying appropriate signal processing techniques.

1.7.4 Noise in Digital Systems

Most noise in digital systems is created by the system itself. Major noise sources include power supply noise, signal return coupling, cross talk, intersymbol interference, and parameter offsets. Lower level noise sources including alpha particles, thermal noise, shot noise, and flicker noise also exist, as explained above. In addition, digital systems impose additional noise due to the nature of their operation. Some of the important noise generated in digital systems is quantization noise and phase jitter noise. Noise generated due to digital operations needs careful attention in the design and implementation of RF communication systems, as it can cause serious errors during data transmissions.

Quantization noise is the result of the difference between the signal and its equivalent quantized value as explained in Section 1.3.

Phase jitter noise is caused by variations in the phase of the carrier as it passes through a communication network. The variations in the phase of the signal cause a smearing in the zero-crossing of the received signal. As a result, severe jitter may result in pulses moving into time slots allocated for neighboring pulses.

1.8 Conclusion

This chapter provides background knowledge on measurements, instruments, sensors, and communication. Essential components of instruments and sensors are explained. Signals generated by sensors have to be handled carefully to extract and transmit the useful information. Wireless instrument communication was introduced and its applications in industrial environments discussed. It was also shown that noise, interference, and distortion can significantly affect communication systems.

2

Wireless Communication

Communication is the transfer of information from one device to another. Modern communication systems involve man-made signals that can be transmitted to different places. The communication signals that carry the information can be in the form of electrical energy, such as currents and voltages, electromagnetic energy, optical energy, or sonic energy. Signals are generated electronically and transmitted via cables and wired lines or wirelessly through electromagnetic radiation in space in the form of radio waves or microwave energy. Signals can also be transmitted by optical or sonic methods.

Communication between instruments takes place via transmission of electrical signals from a source to a sink. The originator of the information is called the source, and the receiving end is the sink. The source converts the original message (e.g., voice, text) into electrical signals that are transmitted to the sink. The source also produces electrical signals suitable for transmission in the selected media—wired, wireless, optical, and so on. The hardware and software that comprises the source is called the transmitter. The task of the transmitter is to process the communication signal into a suitable form for successful transmission along a selected channel. The communication channel is the medium that connects the transmitter to the receiver. The medium can be wires, coaxial cables, fiber-optic cables, or space that carries electromagnetic waves or light waves. On the sink side, a receiver extracts the signal coming from the communication channel and processes it such that the information can be interpreted and understood. The sink converts the electrical signal from the receiver back into the form of the original information, such as voice, text, or data.

This chapter concentrates on the basic principles of wireless communication techniques using electromagnetic waves, particularly at radio frequencies. Wireless communication techniques are being used in a wide range of applications from industrial systems to a variety of consumer products.

The chapter starts by introducing the basic principles of electromagnetic wave propagation. The important elements of electromagnetic radiation, losses, fading, reflection, refraction, and attenuation are discussed. The components necessary for successful RF communication are introduced. Modern communication methods, modulation and multiplexing techniques, fre-

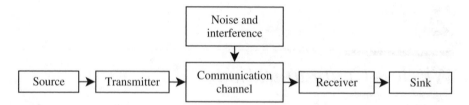

FIGURE 2.1
Block diagram of a simple communication system.

quency spreading, and multiple access methods relevant to wireless sensors and instruments are explained in detail.

2.1 Wireless Communication Principles

In wireless communication systems, information is transmitted from the source to the sink through a communication channel, as shown in Figure 2.1. The information is carried by messages. A message contains all the useful features of the information plus additions such as encryption and protocols. Messages can be in a variety of forms, as long as the source and sink can understand each other. In instrumentation systems, most messages are in the form of continuous, but time-varying quantities, such as the temperature of a process, pressure of a vessel, acceleration of a vehicle, or various chemical variables.

Wireless communication uses electromagnetic waves propagating in free space or a medium. Electromagnetic waves are generated by an antenna that transforms electric power into electromagnetic energy. A receiving antenna picks up the radiated electromagnetic energy and transforms it into electrical signals. The transmitted and received energy are generally expressed in terms of decibels.

2.1.1 The Decibel

The decibel (dB) is a measure of the ratio of the relative power between the output and input of a communication system. Since it is a relative ratio, decibels do not indicate absolute signal levels at some point in time and space. The decibel is defined as

$$10\log\frac{P_{out}}{P_{in}}dB\ .$$ (2.1)

As an example, an amplifier has an input of 2 mW and an output of 5 mW:

$$10\log\frac{P_{out}}{P_{in}} = 10\log\frac{5mW}{2mW} \cong 4dB \ . \tag{2.2}$$

Similarly, if the input is 10 mW and the output is 100 mW, the power ratio is 10, which is 10 dB. An output:input ratio of 100 gives 20 dB, a ratio of 1000 gives 30 dB, and so on.

If the power output is less than the input, the gain of the circuit assumes negative decibels. For an output of 1 mW and an input of 2 mW, the ratio is 0.5 or −3 dB. An output:input ratio of 0.1 is −10 dB, a ratio of 0.01 is −20 dB, and so on.

There are several meanings for 0 dB: 0 dBj = 1 mV, 0 dBk = 1 kW, 0 dBm = 1 mW at 600 ohms, 0 dBv = 1 V, and so on. Giving some examples of the most common decibels in communication engineering, the decibel referred to 1 mW of input is known as the dBm; therefore, 1 mW output for 1 mW input is represented by 0 dBm. In dBm, the input is understood to be 1 mW, thus it is expressed relative to an input of 1 mW. Following this philosophy, an output of 10^2 mW is 20 dBm; similarly, an output of 300 mW is 24 dBm. As mentioned earlier, in both cases, the input power is assumed to be 1 mW. In actual operations, if the input power assumes any value other than 1 mW, then the output power changes proportionately, hence the ratio in dBm remains the same.

Similarly, a decibel referred to 1 W is known as dBW, hence a 1 W output is 0 dBW. An output of 1 mW is −30 dBW, and an output of 20 W is 13 dBW.

The expression of currents and voltages in decibels is slightly different. Assuming that the resistances of the input and output are the same, using the definition of decibels we get

$$10\log\frac{P_{out}}{P_{in}} = 10\log\frac{\left(E_{out}\right)_R^2}{\left(E_{in}\right)_R^2} = 20\log\frac{E_{out}}{E_{in}} \ . \tag{2.3}$$

In radio engineering, circuits are often quoted in dBmV or dBuV. In the case of dBuV, a decibel is relative to a reference signal level of 1 µV. Therefore a signal level of 20 dBuV is actually 10 µV; similarly 0.5 µV is 3 dBuV. Similar results are applicable for dBmV.

2.2 Electromagnetic Wave Propagation

In radio and microwave communication systems, an antenna transforms electric power generated by the circuits into radiating electromagnetic energy. Electromagnetic energy propagates through the media as transverse electromagnetic (TEM) waves. The wave has a transverse electric field (E)

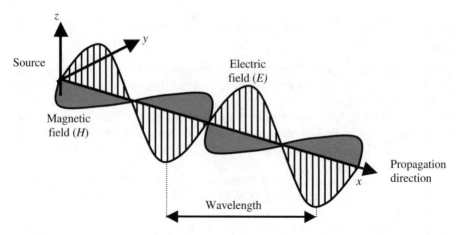

FIGURE 2.2
Propagation of plane waves.

and a transverse magnetic field (*H*) propagating at the same time at right angles to each other and both being at right angles to the direction of propagation. The electric field strength, *E*, is expressed in volts per meter (V/m) and the magnetic field strength, *H*, is in amperes per meter (A/m). The propagation of a TEM wave is illustrated in Figure 2.2.

The wavelength of a propagating wave is the distance that it travels in one complete cycle. The distance traveled in one cycle depends on the speed of the wave in the medium. In free space, the speed of RF electromagnetic waves is equal to the speed of light (approximately 3×10^8 m/s), which can be calculated as

$$\lambda = c/f, \tag{2.4}$$

where λ is the wavelength in meters, *c* is the speed of light, and *f* is the frequency in hertz.

The essential properties of electromagnetic waves are wavelength, frequency, intensity or power density, and direction of polarization. The wavelength is the distance that an electromagnetic wave travels in one complete cycle. The frequency is the number of cycles per second. The intensity is the electric field strength of the wave, expressed in volts per unit distance (e.g., V/m). The power density of the wave indicates the power carried, in units of watts per unit area (e.g., W/m²). Many engineers conduct performance calculations of devices and systems in power densities. The direction of polarization is the direction of the free space propagation of the electric field. If the direction of polarization is horizontal with respect to the surface of the Earth, the wave is said to have horizontal polarization. Other polarizations are possible, such as elliptical polarization or circular polarization. Such waves demonstrate rotating electric field vectors (E-vectors). Circular polarization, for example, is assumed to be composed of two linearly polarized waves that have orthogonal polarization properties that change over time. In such cases, the polarization vectors rotate either left or right as the wave propagates in space. The rotation depends on the phase shift between the two linearly polarized components as +90° or −90°.

2.2.1 Power Aspects of Free-Space Propagating and Link Analysis

The power density of a TEM wave at a point in free space can be expressed in terms of two quantities: the E-field and the characteristic impedance:

$$P_1 = E^2 / Z_0, \qquad (2.5)$$

where P_1 is the power density (W/m²), E is the E-field strength (V/m), and Z_0 is the characteristic impedance, which is equal to 377Ω in free space.

In free space propagation, both the electric and magnetic field intensities decrease directly with the distance from the transmitter. The power intensity decreases as the square of that distance. For example, if a point X is three times the distance from the transmitter as the point Y, the E-field intensity at point X will be one-third of the intensity at point Y. But the power density at point X will be one-ninth of the power density at point Y. This is important in modern wireless systems since the power delivered to the receiver diminishes with the square of the distance, thus limiting the effective range of operation.

The free space propagation model is suitable for predicting the strength of the received signal when the transmitter and receiver have a clear and unobstructed line of sight between them. As the antenna receives the signal, it affects the signal by its own gain. The free space power received by a receiver antenna separated from a radiating transmitter antenna by a distance d is given by the Friis free space equation as

$$P_r(d) = P_T G_T G_R \lambda^2 / (4\pi)^2 d^2 L , \qquad (2.6)$$

where P_T is the transmitted power, $P_r(d)$ is the received power, G_T is the transmitter antenna gain, G_R is the receiver antenna gain, d is the distance between the transmitter and receiver, L is the loss factor not related to propagation (≥ 1), and λ is the wavelength in meters. The loss factor L is made up of miscellaneous components such as transmission line attenuation, filter losses, antenna losses, etc.

The Friis free space model is valid for limited values of d for the far-field electromagnetic wave caused by the transmitting antenna. The region beyond the far-field distance, d_f, may be related to the largest linear dimensions of the transmitter antenna aperture and the carrier wavelength, which can be expressed as

$$d_f = 2D^2/\lambda, \tag{2.7}$$

where D is the largest physical dimension of the antenna.

Communication system link analysis can be conducted by a set of calculations that involves the signal:noise ratio. The equation for noise power at the receiver can be expressed as

$$N = kT_0BF_n, \tag{2.8}$$

where kT_0 is the reference noise level ($= 4 \times 10^{-18}$ W/Hz), B is the receiver bandwidth in hertz, and F_n is the receiver system noise factor.

From equations 2.7 and 2.8 the signal:noise ratio can be calculated as

$$S/N = P_TG_TG_{R\lambda}^2/[(4\pi)^2d^2LkT_0BF_n]. \tag{2.9}$$

2.2.2 Antenna Characteristics

Antennas generate electromagnetic waves to propagate in free space. If the electromagnetic radiation from an antenna is uniform in all directions, it is called an isotropic antenna. An ideal isotropic antenna that propagates waves uniformly in all directions does not exist in practice. It is convenient to use the operation of isotropic antennas as references in determining the operational characteristics of practical antennas. For example, the gain of an antenna in a given direction is a measure of how the power level in that direction compares if an isotropic antenna had been used instead.

The antenna is said to be bilateral if the gain of it is the same as a receiving antenna and a transmitting antenna. In the case of receiving antennas, the effective capture area (also called the effective aperture) may be expressed as

$$A_f = G_R\lambda^2/4\pi, \tag{2.10}$$

where A_f is the effective aperture of the antenna, λ is the wavelength, and G_R is the receiver antenna gain. This formula indicates that the effective aperture, A_f, of an antenna depends on the physical size of the antenna.

2.2.3 Near Field, Far Field, and Fading

An electromagnetic propagation disseminated by a source can be viewed to take place in two separable forms: near-field propagation and far-field propagation. Near-field propagation takes place over distances of less than about one-sixth of the wavelength from the source. The near field is also known as the electric field and the magnetic field. In this region, the propagation wave depends on the characteristics of the source and the circuit impedance that generates the waves. In the near field, reactive energy dominates. Reactive energy can be divided and treated as two separate components: the electric field and the magnetic field.

Far field is propagation of waves at distances greater than one-sixth of the wavelength from the source. Once it occurs, the propagation becomes independent of the characteristics of the source. Far-field propagation assumes constant wave impedance, irrespective of the space and location it is propagating. The impedance is defined as the ratio of the electric field intensity to the magnetic field intensity, at 377Ω in free space, which basically represents the ratio of voltage to current.

Since electromagnetic propagation exhibits different characteristics, the techniques of measurement differ also. In the near-field regions, both E- and H-fields must be measured separately. Far-field strengths are normally measured in terms of E-fields. Alternatively, the far-field strength can be measured in terms of power density (in W/m^2). The power density denotes the amount of radiated power passing through each square meter of a surface perpendicular to the direction away from the source. The peak power density is equal to $E \times H$, and for a sinusoidal source, the average power density is half of this value. In the case of sinusoidal sources, each frequency component must be considered separately. The total average power can be determined as the sum of the average powers for all frequencies. It follows that the total average power is $P = E^2/377$ W.

During measurements of the field strength and intensity of radio communication signals, large variations in the intensity of the field are found over distances with a minimum scale length of half the wavelength, λ. At the smallest scale, in the near-field region, the variations are sensitive to the frequency of the waves and the presence of objects in the path of propagation. Variations depend on the time as well as the shapes and positions of objects, such as the movement of people in the vicinity. With the use of moving receivers, rapid near-field spatial variations (also known as fast fading) can be monitored conveniently. Variations can be explained in terms of the existence of multiple paths as the electromagnetic energy propagates away from the antenna. Variations over a distance greater than half the wavelength show only a gradual variation with frequency; this is called slow fading.

2.2.4 Electrical Field Vector Addition

E-fields are vector quantities that can be described by directions, magnitudes, and phases, and thus they can be treated mathematically as standard vector elements. Two or more E-fields can be added or subtracted in vector form for various reasons; for example, fields from individual elements of a phase array antenna need to be treated as vectors to find the resultant field. Also, if a propagating wave reflects, the vectorial addition or subtraction of direct and reflected waves is necessary to determine the overall spatial field strength. The spatial field strength of propagation depends on factors such as the intensity of the reflections and the phase differences between the propagated and reflected waves.

2.2.5 Free Space Path Loss

The term path loss (also known as free space loss) refers to the spreading loss of a radiated signal between the transmitter and receiver antennas when the gain is equal to one. Path loss represents signal attenuation as a positive quantity and is measured in decibels. Path loss can be expressed as

$$P_L = \lambda^2/(4\pi)^2 d^2, \tag{2.11}$$

where P_L is the free space path loss as a scalar quantity, d is the distance in meters, and λ is the wavelength in meters.

If antenna gains are taken into account, path loss can be expressed as

$$P_L = G_T G_R \lambda^2/(4\pi)^2 d^2. \tag{2.12}$$

Free space path loss increases as the frequency increases. This is because the capture area of the antenna becomes smaller as the distance increases. Path loss also increases rapidly with increasing distance due to the inverse proportionality of distance squared.

2.2.6 Excess Path Loss and Atmospheric Attenuation

Radiated electromagnetic waves propagating in free space are subject to many losses, including diffraction, ionosphere refraction, and scatter propagation losses. Collectively these losses are called the excess path loss.

As in the case of satellites and space explorations, when the electromagnetic waves pass through the atmosphere they attenuate due to atmospheric absorption. At lower frequencies, the additional atmospheric attenuation is small, thus the propagation losses are close to that of free space propagation losses. But at high frequencies, 10 GHz or greater, atmospheric absorption loss becomes excessive. In addition to the normal atmospheric losses, factors such as rain, snow, and fog cause additional losses. These additional losses become fairly large, particularly at 1.3 GHz or greater.

2.2.7 Reflection of Electromagnetic Waves

Electromagnetic waves can reflect from the Earth's surface and other objects. The reflected wave may interfere with the direct wave. If the interference is constructive, the reflected wave is added to the direct wave, thus producing a larger wave at the receiver antenna. If the reflected wave is destructive, it causes a smaller wave than the original direct wave. The constructiveness or destructiveness of interference depends on the relative phase angles of the two interacting signals. The vector addition of a direct signal and a reflected signal is referred to as multipath. Generally multipath is not desirable, but it can be put in a useful form in some situations, as in the case of some indoor wireless systems.

Multipath can be very important in communication systems. For example, in cellular systems, multipath can be a propagation problem that needs careful attention, as there may be a number of propagation paths between the transmitter and the receiver. The desired path from the transmitter to the receiver is a direct path, but other signals will be received due to reflection from the ground and other objects. In cellular systems, the reflected signals may add constructively, resulting in signals greater than a direct signal, which may improve reception quality.

2.2.8 Atmospheric Refraction

An important effect of the atmosphere on electromagnetic wave propagation is refraction. The atmosphere does not have a uniform density and its chemical composition and ionization effects change with altitude. An implication of this is that the propagation velocity of a wave changes as it travels through the atmosphere. As a result, transmitted waves do not propagate in straight lines, but follow curved paths, thus slightly changing their velocity as they pass through different atmospheric conditions.

2.2.9 Diffraction of Electromagnetic Waves

The term diffraction refers to bending of electromagnetic waves by materials along the propagation path. Large objects such as buildings, houses, warehouses, and trees cause diffraction in open spaces (as shown in Figure 2.3).

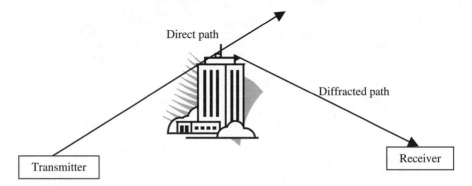

FIGURE 2.3
Diffraction of electromagnetic waves.

Objects such as machinery, columns, walls, and other structures cause diffraction in closed spaces such as plant floors and offices. In both open and closed spaces, the diffraction effect may be useful as a fraction of the energy in a wave bent around an object (e.g., buildings) may be received by a receiver that does not have a direct line of sight from the transmitter. Although there is a loss associated with the diffraction process, the signals received may be strong enough for good reception.

2.2.10 Indoor Propagation of Electromagnetic Waves

The properties of electromagnetic waves in an indoor environment are very important in wireless instruments and networks in factories and plants. The signal transmitted from a fixed base station in an indoor environment reaches the receiver via one or more paths, all slightly delayed in time. The received waves may consist of a line-of-sight signal and other signals reflected or scattered by structures such as outer walls, columns, and so on.

Multipath fading can seriously degrade the performance of an indoor communication system, but the effect can be minimized by proper design of the transmitter and receiver and carefully selecting their locations. In indoor radio communication, a detailed characterization of the spatial properties of signal propagation may be necessary for successful operation of the system.

2.2.11 Frequency Allocation

Electromagnetic radiation has different properties at different frequency bands and can be divided into several groups, as shown in Table 2.1. Some

TABLE 2.1

Frequency bands of electromagnetic radiation

Frequency	Wavelength	Frequency band
3–30 kHz	10^5–10^4 m	VLF (very low frequency)
30–300 kHz	10^4–10^3 m	LF (low frequency)
0.3–3 MHz	10^3–10^2 m	MF (medium frequency)
3–30 MHz	10^2–10 m	HF (high frequency)
30–300 MHz	10–1 m	VHF (very high frequency)
0.3–3 GHz	1–0.1 m	UHF (ultra high frequency)
3–30 GHz	10–1 cm	SHF (super high frequency)
30–300 GHz	1–0.1 cm	EHF (extremely high frequency)
0.3–3 THz	1–0.1 mm	Band 12
1–417 THz	300–0.72 mm	Infrared
417–789 THz	0.72–0.38 mm	Visible light
789–5×10^6	0.38–6×10^{-5} mm	Ultraviolet
3×10^4–3×10^8 THz	100–1×10^{-2} Å	X-rays
$>3 \times 10^7$	<0.1 Å	Gamma rays

of these frequency bands are extensively used for communication purposes and those bands will be discussed briefly.

The very low frequency (VLF) and low frequency (LF) bands have large wavelengths requiring large antennas. The bandwidths of VLF and LF are small, thus limiting their applications. These frequency bands are inefficient for use in communication systems, but they are extensively used in submarine and maritime communication and radio navigation systems.

The medium frequency (MF) band is used in communication involving moderate distances. This band is extensively used in commercial radio broadcasting, maritime and aeronautical communication, and navigation systems. Amateur radio broadcasting is also allowed in this band.

The high frequency (HF) band provides good propagation properties over long distances. This band is used in some fixed and mobile services.

Very high frequency (VHF) and ultra high frequency (UHF) are commonly used in communication systems that require large bandwidths. They can be transmitted and received by small directional antennas, thus adding to the efficient use of these bands. The VHF and UHF bands are used in fixed communication services, ground-to-air communication, television broadcasting, and some mobile services. These bands are particularly efficient in applications where line-of-sight operations are possible.

Super high frequency (SHF) and extremely high frequency (EHF) have very short wavelengths and thus require small antennas. In these bands, the distance of communication is short, and long distance propagation requires a series of repeaters and relay stations. These bands are used for television broadcasting and high-speed data transmission applications.

Frequency bands above EHF are used for communication purposes with fiber optics. Infrared frequencies are used in wireless local area networks.

The frequency allocations for different communication services are provided on a worldwide basis by the International Telecommunication Union (ITU). The allocation for Region 2 (for North and South America) has been adopted by the U.S. Federal Communications Commission (FCC). The FCC is responsible for RF spectrum management within the United States. Individual frequency assignments must be requested by the user and approved by the FCC before any user can transmit at the requested frequency. Other FCC regulations must be followed, such as the maximum transmitted power, out-of-band harmonics, spurious signal levels, and so on.

2.3   RF Components

Modern electronic communication is largely based on digital systems. The major functions and components of a typical transmitter for a digital communication system are shown in Figure 2.4. This figure is important because

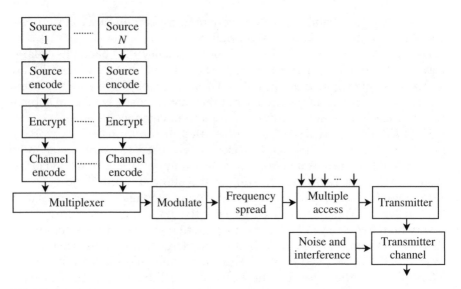

FIGURE 2.4
Components of a typical wireless digital transmitter.

most of the wireless systems discussed in this book have similar architectures in their construction and operation.

The functions of the components of a digital transmitter (Figure 2.4) are as follows: The source converts a signal (e.g., voice, measurement result from an instrument, etc.) into electrical signals. The output signals of the source are compressed through source coding, since there is usually redundancy in the original signal. Source coding also shapes the signal to make it suitable for transmission within the bandwidth of the communication system. Encryption encodes the signal into a complex form to prevent interception and interpretation of the signal by third parties or unauthorized recipients. Channel encoding reformats the signal to prepare it for any anticipated noise and distortion that might take place in the selected channel during transmission. Once information on the characteristics of the channel encoding of the transmitter is known by the receiver, it can take corrective action for errors that might have occurred during transmission. If there is more than one transmitted signal, multiplexers are used to merge a number of signals together to be transmitted on the same channel. Multiplexers increase efficiency, since channel bandwidth is normally greater than the bandwidth of any one of the transmitted signals. Multiplexed signals are then modulated and frequency spread. Frequency spreading has many advantages; for example, the signature of the transmitted signal is reduced so that the probability of interception and interpretation is low. After frequency spreading, multiple access may be allowed by sharing the channel with other users. Once the signals are modulated, they are then ready for transmission across the channel.

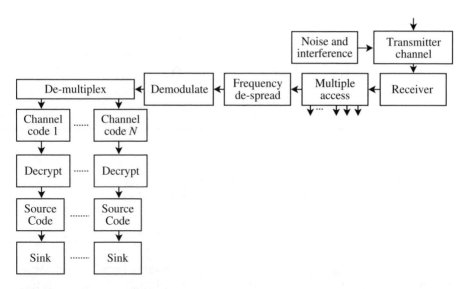

FIGURE 2.5
Components of a typical wireless digital receiver.

The receiver has components that extract the intended information by despreading, demodulating, and demultiplexing the received signals coming from the transmitter. The major functions and components of a typical receiver are illustrated in Figure 2.5.

Impedance matching between components is significant because if there is an impedance mismatch in the system, some energy intended for transmission reflects back to the source, causing standing waves. The amount of energy that reflects back depends on the frequency of the signal and the degree of mismatch. Standing waves are expressed in terms of the voltage standing wave ratio (VSWR). The VSWR is a unitless ratio ranging from one to infinity, mainly showing the amount of reflected energy that exists in a device. A value of one for the VSWR indicates that all the energy passes through without any reflection, while higher values indicate an increase in the amount of reflection.

Clearly the architecture and components of communication systems vary. However, there are a number of components that perform similar functions and are common in most RF communication systems, including amplifiers, attenuators, filters, oscillators, multipliers, mixers, modulators, demodulators, multiplexers, and antennas. Other important components of modern systems are phase detectors and phase shifters, delay lines, duplexers and diplexers, isolators and circulators, up converters and down converters, power dividers and power combiners, transformers, electromagnetic interference (EMI) and radio frequency interference (RFI) filters, gain controllers, transceivers, and wireless modems. A block diagram of the components of a typical receiver system is illustrated in Figure 2.6. Some of the common

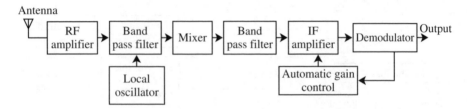

FIGURE 2.6
Components of a typical receiver system.

components will be explained next to provide background information for designers and users of wireless instruments and networks.

2.3.1 Amplifiers

Amplifiers are electronic devices that produce output signals with larger amplitudes compared to the input signals. In an ideal amplifier, the output signal should resemble the wave shape and frequency of the input signal, except the amplitude should be increased by a set proportion called gain. In radio engineering, many different amplifiers are designed and used. Types of RF amplifiers include low noise amplifiers, power amplifiers, pulse amplifiers, bidirectional amplifiers, multicarrier amplifiers, buffer amplifiers, and limiting amplifiers. The input signal of an amplifier can be a current, a voltage, or any other signal, and the output signal is usually of the same nature as the input.

Specifications for RF amplifiers include three main characteristics: the frequency range, the gain, and the output power. In voltage amplifiers, the gain is the ratio of the output voltage to the input voltage of the signal. In current amplifiers, the gain is expressed as the ratio of output current to input current. Similarly, in power amplifiers, the ratio of output power to input power is the power gain. Normally gains are expressed in decibels (e.g., $G_{dB} = 20 \times \log(V_o/V_i)$ for voltage gain or $G_{dB} = 10 \times \log(P_o/P_i)$ for power gain. The output power is determined by the signal power at the output of the amplifier under specific conditions such as temperature, load, VSWR, or supply voltage. The output power is expressed in dBm, milliwatts (mW), or watts (W).

The electrical characteristics of RF amplifiers include nominal operating voltage, current, and impedance. Additional performance specifications include the noise figure and the input VSWR. The noise figure is the ratio (in dB) of the signal:noise ratio at the input and the signal:noise ratio at the output of a component. Thus the noise figure indicates a measure of the amount of noise added to the signal under the normal operations. Clearly a low noise figure indicates better performance. The noise figure sets the lower limit of the dynamic range of an amplifier.

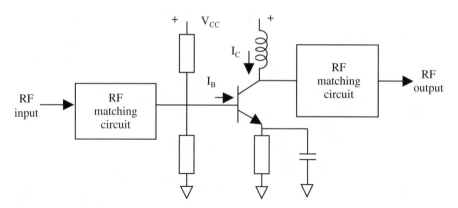

FIGURE 2.7
A typical class A RF amplifier.

Radio frequency amplifiers are essential building blocks of RF systems and they find applications as transmitter RF amplifiers, receiver RF amplifiers, intermediate frequency (IF) amplifiers, low frequency (LF) amplifiers, transmitter output power amplifiers, and buffer amplifiers.

Radio frequency amplifiers are used on both transmitter and receiver sides and are an important part of radio systems. They are used for all RFs, including VHF and higher frequency transmitters and receivers. In all applications, special care must be taken to design and construct low noise RF amplifiers by using carefully selected bipolar junction transistors (BJTs) and associated components or field effect transistors (FETs). Depending on the application, amplifiers can be designed and constructed in different types, such as class A, class B, class C, push-pull amplifiers, and so on.

Class A amplifiers, as shown in Figure 2.7, are generally used for low noise amplification of small signals. They have good linearity, low distortion, broad bandwidth, good noise figure, low band-pass ripple for all output levels, medium output power capability, and stable phase and gain, but have poor efficiency, more heat dissipation, and are large in size.

Class B and class AB amplifiers are used for large-scale and high-power amplifications. Class C amplifiers have non linearity that may not be acceptable in some applications. In all cases, in the design of amplifiers, impedance matching between the circuits is important. Impedance matching is often realized by lumped capacitive-inductive circuits known as LC circuits. In the ideal case, the signal input of an RF amplifier should result in an increase in amplitude without changing the frequency components or introducing any other form of distortion. Nevertheless, in practice there will be some distortions in the output signals due to the nonlinear operation of transistors and other components.

Intermediate frequency amplifiers are used for providing the necessary gains for receiver circuits. They are preferred because it is easier to construct narrowband band-pass filters at IFs. These amplifiers usually include automatic gain controllers (AGCs).

Low frequency amplifiers are often used in RF systems, mainly after the modulators in the transmitters and before the demodulators in receivers. In many applications, RC or transformer-coupled class A amplifiers are preferred as LF amplifiers. In come cases, class AB push-pull amplifiers are preferred.

Transmitter output power amplifiers are essential amplifiers for delivering the necessary power to the transmitter. These amplifiers must be carefully selected or designed so they do not violate FCC limitations of maximum power levels that can be transmitted at a certain frequency. They are constructed from BJT or FET transistorized power amplifiers, often as single-ended or push-pull amplifiers. In many applications involving high power transmission, a number of cascaded amplifiers may be connected in chains. Chain amplifiers may use frequency multipliers to achieve the required operational specifications.

2.3.2   Attenuators

Attenuators are circuits that reduce the power level of a signal by a set amount, ideally without frequency change, distortion, or reflection. Basically the output signal is reduced with respect to the input signal. Attenuators used in radio systems can be fixed or variable. A fixed attenuator reduces the power level of a signal by a predetermined amount, but it has to be designed carefully with respect to the input and output characteristics of the signals involved. A variable attenuator has a specified attenuation range that makes it adaptable to different circuit conditions.

2.3.3   Filters

Filters are one of the most important components of RF circuits. All types of filters are used in RF applications, including low-pass filters, high-pass filters, band-pass filters, band-stop filters, and notch filters. A low-pass filter passes low frequencies and rejects high frequencies. A high-pass filter passes high frequencies while rejecting low ones. A band-pass filter allows the signals to pass through within the desired band. A band-stop filter blocks specified frequencies while passing frequencies outside that band. Notch filters are narrowband band-pass or band-stop filters.

Performance specifications of RF filters include considerations such as the cutoff frequency, bandwidth, ripple, insertion loss, and VSWR. For low-pass filters, there is one specified frequency—the cutoff frequency—above which the frequencies are blocked. For high-pass filters, the specified frequency refers to the cut-on frequency, the higher frequencies that are allowed to pass. For band-pass and band-stop filters, an important specified frequency is the center frequency in the middle of the bandwidth. Specified lower and upper frequencies determine the range of the bandwidth. Ripple is the peak-to-peak variation in the pass band response. Ripple is expressed in decibels and is a function of the degree of matching of the devices in the system. Insertion loss is the total RF power transmission loss resulting from the insertion of a device in a transmission line. It is defined as the ratio of signal power at the output of the inserted device to the input signal power at the input of the inserted device.

2.3.4 Oscillators

The generation of different frequency signals is an important part of transmitters and receivers in communication systems. Different frequency signals are generated by oscillator circuits using suitable components such as transistor feedback oscillators. These types of oscillators are basically transistor amplifiers with positive feedback. The frequency of oscillation is determined by the resonant circuit, usually located in the feedback circuit. There are many types of oscillators that are used in RF applications: LC oscillators, crystal oscillators, dielectric resonator oscillators (DROs), yttrium-iron-garnet (YIG) resonator oscillators, and ceramic resonators. Only a few of these oscillators will be discussed here.

LC oscillators are frequently used in RF communication systems. There are many different types of LC oscillators, such as Colpitts oscillators, Clapp oscillators, tuned-input tuned-output junction FET (JFET) oscillators, differential pair oscillators, among others.

Quartz crystal oscillators are one of the most common oscillators used in RF systems. Quartz crystal oscillators provide steady frequencies with high accuracy. They can be tuned to precise frequencies by using external trimming capacitors.

Dielectric resonator oscillators make use of the properties of high-Q dielectric materials such as resonators. They provide stable frequencies and exhibit good phase noise properties. YIG resonator oscillators are used in UHF and microwave applications. They are an essential part of mixers and frequency converters.

2.3.5 Frequency Multipliers

Frequency multipliers are essential components of RF communication hardware. Multipliers produce larger frequency output signals than the input signal frequency by a predetermined amount. Although they are nonlinear devices, most frequency multipliers are able to eliminate undesired harmonics from the output signal, thus acting like a filter. There are two types of frequency multipliers: active multipliers and passive multipliers. An active multiplier produces an output signal with a higher power level than the input signal power level. The difference in power level (expressed in dB) between the output signal and the input signal is called the conversion gain.

The conversion gain is a positive number. A passive multiplier produces an output signal with a power level smaller than the input signal power level. The difference in power level between the output signal and the input signal is called the conversion loss. The conversion loss is a negative number, but generally it is specified as an absolute value.

Frequency multipliers can be produced by using diodes or transistors. Transistor-based frequency multipliers are normally made from low power transistors operating in class C. A circuit of a class C multiplier is shown in Figure 2.8. In this figure, the tuned circuit is resonant at $n \times f$, where f is the frequency of the input voltage and n is usually 2 or 3. Higher multiples may be used, but the output may be very low. The tuned circuit consists of a small coil and variable capacitor or a slug tuned coil and fixed capacitor.

Diode multipliers are also fairly common because of their simplicity and efficiency. Two common diode multipliers are the varactor and the step recovery diode frequency multiplier.

2.3.6   Mixers

Mixers have two input signals at different frequencies and two outputs. Their function is to convert RF power of one frequency into power at the other frequency. The signal of the power to be shifted is applied at the RF input. The signal required to carry the power (usually obtained from a local oscillator [LO], thus it contains low power) is applied to the LO port. The frequency sum and the frequency difference of the two signals are available at the IF port.

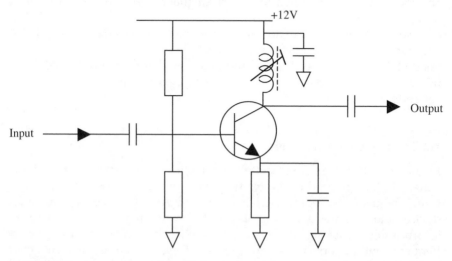

FIGURE 2.8
Basic circuit of a transistor frequency multiplier.

Performance specifications of mixers include RF range, LO frequency range, IF range, conversion loss, and LO power. The RF range is the range of input frequencies that the mixer can handle without errors. The LO frequency range is the frequency range of the local oscillator. The IF range represents the selection of signals that will be subjected to further signal processing for transmission, in the case of transmitters, or signal extraction, in the case of receivers. For example, in the case of superheterodyne receivers, the incoming signal is mixed with a LO signal to process the sum or difference of the two signals that represent the IFs of the receiver. Conversion loss is defined as the loss of power between the RF input signal and the IF output signal during frequency translation. It is calculated in decibels as the ratio of the IF output power to the RF input power. The conversion loss is a measure of the efficiency of the mixer. Conversion loss measurements are normally conducted by observing the mixer input/output characteristics on a 50Ω system having a fixed LO signal power. The LO power is the required power level, in dBm, that must be supplied by the local oscillator in order to successfully drive the mixer diodes. The LO power is a very important factor in determining the dynamic range of a mixer.

There are many common mixer types that are commercially available or can be constructed: single balanced mixer, double balanced mixer, triple balanced mixer, quadrature mixer, and image reject mixer. The majority of these mixers are made from diodes. There are also different types of diode mixers used in modern RF systems, such as the single-ended diode mixers, balanced diode mixers, double balanced diode mixers, and triple balanced diode mixers. Transistor mixers that are made either from BJTs or FETs are also used. Transistor mixers also come in different ranges as single-ended mixers, balanced mixers, and double balanced mixers. One important advantage of transistor mixers is that they can provide conversion gain rather than conversion loss. FET mixers are known to produce less intermodulation and cross-modulation distortion, hence they are preferred in high frequency operations.

In the application of RF mixers, isolation is important. Isolation may be defined as the ratio, in decibels, of the power level applied at one port of the mixer to the resulting power level at the same frequency appearing at another port. Isolation is a measure of the internal circuit balance for the two inputs of the mixer and it indicates the level of attenuation of a signal when supplied into a specific port, but its effect is observed at a different port. Clearly, when the isolation is extreme, the amount of "leakage" or "feedthrough" between the mixer ports will be small.

2.3.7 Modulators and Detectors

Modulation is an essential process in communication systems. A modulator changes certain characteristics of a signal such as amplitude, frequency, or phase, and prepares the carrier signal for transmission. Modulators are con-

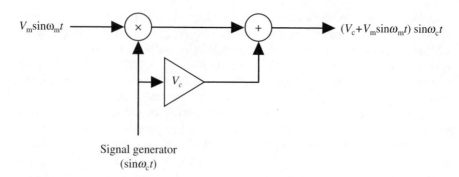

FIGURE 2.9
An AM product modulator.

figured using different technologies, depending on the power and frequency of the signal to be transmitted.

Modulation can be performed by analog or digital methods. Analog modulation is an older, but well-established technology, and is largely used in analog communication systems. Digital modulation, on the other hand, is a new technology that is developing quickly as more and more digital communication systems are developed. While new modulation techniques continue to appear, certain techniques have matured over the years and are extensively employed in many applications.

Analog modulation occurs in three different forms: amplitude modulation (AM), frequency modulation (FM), and phase modulation (PM).

Amplitude modulators are commonly made from a class C amplifier using plate-modulated, grid-modulated, or collector-modulated circuits. There are many other types of conventional amplitude modulators, including FET gate-biased AM systems and FET push-pull modulators. Also, various types of sideband modulators are available in the form of single-sideband modulators, double-sideband modulators, vestigial-sideband modulators, etc. As an example, a product modulator is illustrated in Figure 2.9. This modulator operates on multiplication and addition of the carrier signal and the modulated signal. Signal multiplication at higher frequencies can be accomplished by other methods, such as with the square law modulator shown in Figure 2.10.

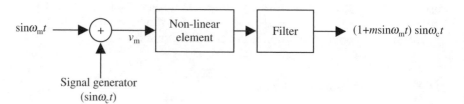

FIGURE 2.10
An AM square law modulator.

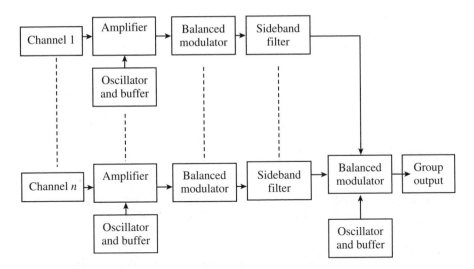

FIGURE 2.11
A basic n-channel FDM.

Frequency modulators use many signals in parallel by using different frequencies for each channel. Frequency modulated signals are often produced by amplifier chains. The modulation can be accomplished either by direct variation of the frequency of an oscillator or indirectly by phase modulation. Thus there are two basic methods for generating an FM signal: the direct method and the indirect method.

In the direct method, the carrier frequency is directly varied in accordance with the input modulating signal. A voltage controlled oscillator (VCO) is used to vary the frequency of the carrier signal in accordance with the variations in the baseband signal amplitude.

In the indirect method, a narrowband FM signal is generated using a balanced modulator and frequency multiplication is used to increase both the frequency deviation and the carrier frequency to the specified level.

Figure 2.11 illustrates a simplified block diagram of a frequency division multiplexing (FDM) obtained using FM modulation for each channel. In this system, the signals are sent using suppressed carrier modulation and each channel has its own amplifier, oscillator, balanced modulator, and sideband filter. Channels are grouped together by adders and group filters; groups can be combined to form supergroups; supergroups form master groups; and so on.

In digital RF systems, the carrier is modulated by a digital baseband signal; this offers many advantages over analogue modulation and is widely used in modern wireless systems. The digital counterparts of analog modulators for AM, PM, and FM are called amplitude shift keying (ASK), phase shift keying (PSK), and frequency shift keying (FSK), respectively.

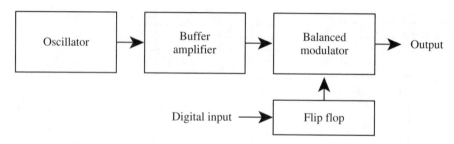

FIGURE 2.12
Block diagram of a BPSK modulator.

Phase shift keying modulators come in various forms using a host of technologies. Typical PSK modulators are binary PSK (BPSK), quaternary PSK (QPSK), and others.

In BPSK, the signal is keyed between two phase states 180° apart. A balanced BPSK modulator is illustrated in Figure 2.12. The amplitude modulator is controlled by a logic circuit and the output of the balanced modulator is normally fed to an amplifier.

Quaternary phase shift keying involves four phase states, with each state separated by 90°. Each pulse represents two bits of information (00, 01, 10, or 11). An important variation of QPSK is $\pi/4$-DGPSK modulation. This is a differential four-state PM technique in which the phase reference is derived from the previous RF pulse. Information is sent by changes in phase from pulse to pulse rather than in absolute phase. The $\pi/4$ indicates that the minimum phase shift with respect to the reference is 45°.

Eight-state phase shift keying (8-PSK) involves the use of eight phase states with the phase states separated by 45°. Each pulse represents three bits of information (001, 001, 010, 011, 100, 101,110, or 111).

Sixteen-quadrature phase shift keying (16-PSK) provides four-bit pulse capability. An improved version of 16-PSK is 16-QAM (16-quadrature amplitude modulation). 16-QAM uses four states and four possible amplitudes. This 4×4 combination results in 16 possible modulation states.

Frequency shift keying modulators are mainly used in digital data transmission and operate as standard modulators except that only two frequency states are used.

All these methods are revisited and detailed in Section 2.5.

2.3.8 Demodulators

Demodulators are used on the receiver side and are constructed to decode the information coming from the transmitter. For successful information transmission and reception, the demodulator on the receiver and the modulator on the transmitter must be a matching pair. Consequently, depending on the modulator, the demodulator can be amplitude, frequency, or phase demodulators.

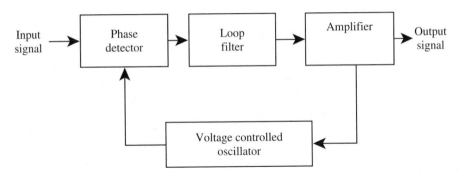

FIGURE 2.13
A PLL FM detector.

Amplitude demodulators convert an amplitude modulated RF signal into an audio, video, or pulse signal of the same original form as the modulated signal. An important type of AM demodulator is called an envelope detector. Improved versions of envelope detectors include volume control and automatic gain control.

Another form of amplitude demodulator is the product detector, which consists of a balanced demodulator or balanced mixer circuit and filters. Product detectors are used to demodulate single-sideband (SSB) or double-sideband (DSB) signals. They are made from either diodes or transistors.

Frequency demodulators, also known as FM detectors, use diodes, transformers, discriminators, rectifiers, limiters, and filters to demodulate the received signals. A phase-locked loop (PLL) FM detector is shown in Figure 2.13. In PLL detectors, as the frequency changes, the error signal that is needed to track the frequency changes proportionately, thus making the error signal a measure of the received signal frequency.

Phase detectors use filters, power splitters, couplers, and mixers, as shown in Figure 2.14. The structures of phase detectors are different depending on the types of signals they handle (e.g., BPSK or QPSK). In the case of BPSK detectors, double balanced diode mixers are often preferred. In this case, one of the inputs of the mixer is the phase-modulated carrier frequency from an IF amplifier and the other input is the reference carrier frequency from an

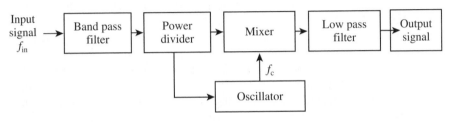

FIGURE 2.14
A typical BPSK phase detector.

oscillator. Consequently the frequency of the output signal of the balanced mixer is either $f_{out} = f_{in} - f_c$ or $f_{out} = f_c - f_{in}$. The correct frequency and phase angle can be obtained by suitable synchronization techniques.

Quaternary phase shift keying demodulation, on the other hand, provides two-bits per pulse. QPSK demodulators are made from filters, three-way power splitters, balanced mixers, quadrature phase couplers, adders, and timing circuits. In some cases, additional power splitters are used in the output stages to recover the carrier frequency. The timing circuits are used to determine the binary states for the signal as one or zero. Timing circuits also provide necessary time delays for conversion of parallel signals into serial forms.

In the demodulation process, if the carrier of the received signal and the oscillator of the receiver requires phase synchronization, this method is called coherent detection. Coherent detectors, also known as correlators, provide a high signal:noise ratio and hence the probability of error becomes low. Coherent detectors require that the phase of the local oscillator in the receiver must have a consistent relation with the carrier of the received signal. Called "carrier recovery," establishing a high degree of agreement in phases requires complex circuits, and is difficult at low signal levels and if there is significant interference and signal fading.

In contrast, some modulated waveforms can be demodulated using non-coherent detection. Noncoherent detection is based on a series of filters to determine received frequencies and to eliminate interference. Noncoherent detectors are used more often because of their simplicity.

2.3.9 Multiplexers

In traditional communication systems, many different signals (e.g., data, image, voice) are transmitted over a single communication link by two generic forms of multiplexing: frequency division multiplexing (FDM) and time division multiplexing (TDM). However, in recent years code division multiplexing (CDM) has been added to the list and is extensively used in digital systems. In FDM, several channels are stacked together so that each channel occupies a unique portion of the frequency spectrum. In TDM, time is shared between channels. FDM and TDM are explained here and CDM is discussed in Section 2.4.

A simplified system block diagram for a FDM transmitter is shown in Figure 2.15. Many different inputs are fed to each subchannel for modulation. Oscillators are used to provide selected frequencies for each channel. The output of each modulator within a particular group is added to pilot channel oscillator signals using RF add circuits. The combined signal feeds a mixer circuit that acts as a frequency up-converter.

In the next stage, the outputs from all the different groups of modulators are combined together. Since the mixers associated with each group use different frequencies, the signal from each group is distinguishable from the

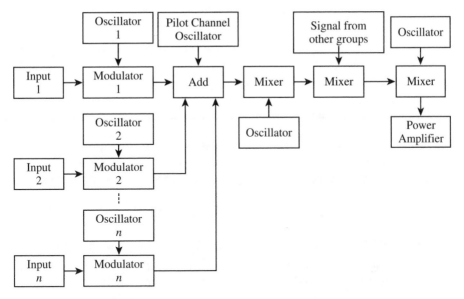

FIGURE 2.15
A simplified FDM system block diagram.

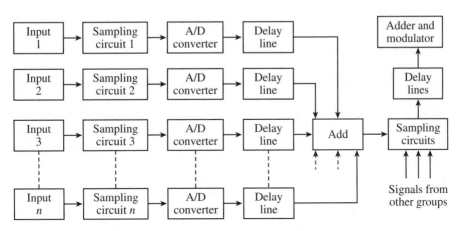

FIGURE 2.16
A typical TDM system.

other signals. This helps in the recovery of each individual signal on the receiving side. A power amplifier chain is used to increase the power level of the combined signal to the desired output level for transmission. The power amplifiers connect the overall signal to the antenna of the transmitter.

A typical TDM is illustrated in Figure 2.16. In a typical TDM, analog inputs are fed to sampling circuits to be converted into digital forms. The task of delay lines is to space data in time. Delay line data are added to be reorganized again in the output stages. The output stage contains other components such as mixers, oscillators, power amplifiers, and the transmitter antenna.

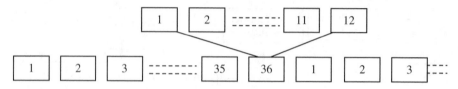

FIGURE 2.17
Time slot assignment in TDM.

The concept of time sharing of a channel is shown in Figure 2.17. In this particular case, there are 36 major time slots and 12 minor time slots in each major time slot. As a result, there are 432 time slots available for data transmission.

2.3.10 Antennas

Antennas are devices that are used to receive or radiate electromagnetic energy. They can be viewed as transducers that convert electrical energy into electromagnetic form or vice versa. Antennas have two basic functions: the receiving antenna intercepts electromagnetic energy and transforms it into electrical energy for processing by appropriate electronic equipment, and the transmitting antenna converts electrical energy from electronic equipment into electromagnetic energy. Electromagnetic energy propagates in space as an electromagnetic field as explained in Section 2.2.

A transmitting antenna is a transducer that transforms electric power into electromagnetic radiation. If the electromagnetic radiation is uniform in all directions, the antenna is called an isotropic antenna. An isotropic antenna is said to be omnidirectional and radiates equal power in all directions in a horizontal plane. Conversely, directional antennas radiate power in a particular direction. If a directional antenna radiates electromagnetic energy in one direction it is said to be unidirectional; if it radiates in two directions it is bidirectional.

There are many different types of antennas, depending on their structure, shape, construction, and material. A thin-wire monopole antenna is commonly used in RF systems. The operating principles of these antennas are similar to coaxial cable. Monopole antennas are constructed in various forms, including wideband, conical, large-size, and electrically small monopole antennas. Monopole antennas are generally fed by coaxial transmission lines. The impedance matching between the cable and antenna is achieved by lumped constant inductor capacitor circuits at low frequencies and stub matching circuits at high frequencies.

Whip antennas resemble monopole antennas, but they are made from flexible rods. They are basically capacitive antennas or electric doublets that are carefully insulated from the structure and supported on a base insulator. The length of whip antennas is usually one-tenth to five-eighths of the wavelength of the radiated electromagnetic waves. Whip antennas are probably the most widely used type in mobile communication systems. Normal mode helix antennas are probably the second most common type.

Microstrip antennas are popular for their low profile and ease of fabrication on printed circuit boards. They can be configured in an unlimited number of shapes and sizes and can conform to odd surface profiles. In addition, they can be constructed using special dielectric materials that fit in small packages. The major drawback is their low efficiency.

Thin-wire dipole antennas are one of the most commonly used antenna types in communication systems. Variations operating on the same principles include electrically small dipole antennas, fat dipole antennas, folded dipole antennas, and turnstile antennas.

There are many other types of antennas, including Yagi, sleeve, loop, helical, spiral, log periodic, slot, notch, horn, lens, and reflector type antennas. In addition, each type may have many variations in their shapes and operating principles.

Antennas can be used in arrays. Array antennas are a collection of various antenna elements, such as dipole, slot, loop, microstrip, etc. Two of the common antennas applied in arrays are Yagi-Uda and log periodic antennas. Yagi antenna arrays consist of three or more half-wave elements, one driver, one reflector, and one or more directors. Yagi arrays offer high directivity and gain. Log periodic arrays are multielement, unidirectional, narrow-beam antennas that have impedance and radiation characteristics that function as a logarithmic function of the excitation frequency. Consequently the length and spacing of the elements in log periodic antennas increase logarithmically from one end to the other. Log periodic antennas are often used in broadband applications.

Panel or sector antennas are another type of array antenna that have multiple elements to radiate or receive signals from different sectors of a cell. Panel or sector antennas usually have flat panel shapes and are enclosed in plastic to protect their elements from each other or other metals, and from other electronic components.

In communication systems, at frequencies less than 3 GHz, many different types of antennas are used, including monopole, dipole, and array antennas. Dish antennas are generally used at microwave frequencies, approximately 3 GHz or greater. In long distance transmissions, satellite communication, and other high bandwidth microwave systems, the most common type of antenna is the dish antenna. For data transmission, commonly available antennas include omnidirectional, directional, array, microstrip, panel, and monopole antennas (e.g., whip antennas).

2.3.11 Phase Detectors and Phase Shifters

Phase detectors are circuits that provide a direct current (DC) output voltage proportional to the phase difference between two same-frequency RF signals with a phase difference between the two. A simple way of obtaining phase

detectors is the use of double balanced mixers. These mixers can be configured as phase detectors by supplying the two RF signals to the local oscillator
and RF input terminals. The DC output of the phase detector (mixer) can be
observed at the IF output terminal.

A phase shifter is a two-port device that alters the phase of a signal in
response to an external signal. There are two basic types of phase shifters:
analog and digital. Analog phase shifters change the output phase by means
of an analog signal. Digital phase shifters use digital signals to change the
output phase. Although there are many different types of phase shifters in
both the analog and digital categories, the variable phase shifter is the most
commonly used type, in which the phase of a signal can be changed by
applying a variable control signal.

2.3.12 Power Dividers and Power Combiners

Power dividers deliver multiple outputs with equal phases and amplitudes
from a single input signal. Power dividers are reciprocal devices that can
be used as power combiners simply by using their output ports as input
ports. Power dividers divide the input signal in two-, three-, four-, five-,
six-, seven-, and eight-way forms.

Power dividers are classified as 0°, 90° (90° hybrid), and 180° dividers (180°
hybrid). A 0° power divider splits an input signal into two or more signals
that are theoretically equal in both amplitude and phase. In hybrid varieties,
the signals are split into different phases, but with equal amplitudes. A 90°
hybrid power divider splits an input signal into two equal amplitude output
signals, but 90° out of phase. A 180° hybrid divider splits an input signal
into two in-phase signals of equal amplitude when an input signal is applied
into one of its two input ports. When the same signal is applied to two input
ports, it produces two equal amplitude signals that are 180° out of phase.
All these power splitters can be used as power combiners as well.

Performance specifications for power dividers include frequency range,
insertion loss, isolation, and RF connector type. The insertion loss, in decibels, is defined as the measured loss through the device excluding the power
division factor. The power division factor is calculated as the ratio of power
output to power input. Isolation is defined as the isolation, in decibels,
between any set of output ports. It is calculated as the ratio of the power of
one output port to the power at any other output port, with matched termi-
nations of all other ports. Common package types for power dividers include
surface mount, flat pack, through-hole or plug-in, and waveguide assembly
technologies.

2.3.13 RF Transformers

Transformers are devices that transfer electrical energy from one circuit to
another by electromagnetic induction. RF transformers are used for increas-

ing or decreasing the voltage. The increase or decrease in the voltage depends on the turns ratio of the secondary and primary windings. If the turns ratio is greater than one, the device acts as a step-up transformer, indicating that the amplitude of the output voltage is larger than the input voltage. If the turns ratio is less one, the device acts as a step-down transformer. RF transformers are used primarily in low-power circuits for impedance matching purposes to achieve maximum power transfer. Transformers provides DC isolation between the input and output circuits while maintaining alternating current (AC) continuity.

Performance specifications for RF transformers are the operating frequency range, 3 dB bandwidth, amplitude unbalance, phase unbalance, impedance ratio, insertion loss, rated DC current, DC resistance, and operating temperature. The 3 dB bandwidth is the frequency range for over which the insertion loss is less than 3 dB referenced to the midband insertion loss. Amplitude balance is the absolute difference in signal amplitude between each output of a center-tapped transformer relative to the position of the center tap. Phase balance, in degrees, refers to the difference in signal phase between each output of a center-tapped transformer relative to the center tap. The impedance ratio is the square of the turn ratio. Insertion loss, in decibels, is the measured loss through the device determined as the ratio of the power output to power input. The rated DC current is the level of continuous DC that can be passed through the transformer windings without causing overheating or damage.

Other specifications for RF transformers include transformer type, package type, packing method, center tap options, and connector type. Typical transformer types are telecom, current sense, power, and Balun transformers. Package types include surface mount, flat pack, through-hole, and plug-in. Packaging methods are tape reel, tray, tube, and bulk pack. Center tap options can be primary side only, secondary side only, and both sides.

2.3.14 EMI and RFI Filters

The term electromagnetic interference (EMI) is interchangeably used with radio frequency interference (RFI). EMI refers to the interference of electromagnetic energy at all possible frequencies, while RFI refers exclusively to the radio frequency range. Both EMI and RFI can be eliminated or minimized with the use of filters. EMI and RFI filters consist of circuits or devices that contain series inductive load-bearing and parallel capacitive non-load-bearing components. Non-load-bearing components provide low impedance paths around the protected circuit for high frequency noise. Filters can also attenuate impulses, since an impulse contains high frequency components, as can be verified by Fourier analysis.

Important specifications for EMI and RFI filters are the rated voltage, rated current, and insertion loss. Insertion loss is a statement of the attenuation characteristics of the filter, in decibels. It indicates the ratio of noise that would get through the circuit without the filter to the noise level that gets through when the filter is installed. Other important specifications are the operating temperature and the temperature range.

2.3.15 Other Components

Delay lines are devices used to slow a signal by a preset time interval. There are two basic types: active delay lines and passive delay lines. Active delay lines are built primarily with digital components and are applied in delaying digital signals. Passive delay lines are built with analog components, but they can delay both analog and digital signals.

Duplexers are devices that allow the use of a single antenna to receive and transmit information. The transmitter operating on one frequency uses the same antenna as the receiver operating on a different frequency. The task of the duplexer is coordination of the time sharing of the transmission and reception activities with a minimum of interaction and degradation of the two RF signals.

Diplexers, on the other hand, are like multiplexers. They are three-port, frequency-dependent devices that are used as a separator or a combiner of different signals. A typical diplexer consists of two fixed tuned band-pass filters that share a common port. The common port and the output of the two filters form the three terminals of the diplexer. Signals applied to the common port are separated in accordance with the pass band frequencies of the filters.

Radio frequency duplexers and diplexers include surface mount, flat pack, through-hole, plug-in, and waveguide assembly techniques. When using RF diplexers and duplexers, important performance specifications to consider are signal rejection of the receiver over transmission signal and vice versa, insertion loss, and VSWR.

Isolators and circulators are two passive devices that are used for controlling the propagation of an RF signal. An isolator permits the signal to pass in one direction while providing high isolation against the reflected signal coming in the reverse direction. A circulator consists of three or more ports that allow the signal entering a port to be transferred to the port adjacent to it in either a clockwise or counterclockwise direction.

Up-converters and down-converters are devices that are used for changing the frequency of an RF signal on a large frequency scale.

2.3.16 RF Transceivers

Transceivers are electronic devices that are used for transmitting and receiving RF signals. For transmission, the transmitter performs all necessary RF signal processing functions, such as modulation, up-conversion, and power amplification, before supplying the signal to the antenna. RF transmitter architecture does not have much flexibility and they are constructed in only a few distinct forms. However, the receiver can operate in different ways, such as heterodyne, homodyne, image-reject, digital IF, subsampling principles, etc.

Heterodyne receivers translate the frequency band of the received signal to much lower frequencies to relax the Q-factor required by the channel select filter. The translation is carried out by a mixer, often referred to as down-conversion. In heterodyne converter design, the IF, which is the center of the down-converted band, is a critical parameter.

Homodyne receivers convert the incoming signal to the carrier frequency. Channel selection requires only low-pass filters with relatively sharp cutoff characteristics.

Image-reject receivers process the signal and the image differently by allowing cancellation of the image by its negated replica. Distinction between the signal and image is possible since they lie on different sides of the local oscillator frequency. There are many different types of image-reject receivers, such as Hartley architecture receivers and Weaver architecture receivers.

Digital IF receivers operate on dual IF heterodyne principles at low frequencies. They need additional mixer and filter circuits.

Subsampling receivers sample RF input at lower rates since the narrowband signals exhibit only a small change from one carrier cycle to the next. Sampling receivers may suffer from aliasing and noise problems.

Important performance specifications for a transceiver are frequency range, sensitivity, output power, and data rate. The frequency range is the operating frequency of the transceiver. The sensitivity is normally taken as the minimum input signal required to produce a specified output signal within a specified signal:noise ratio. The output power is the maximum signal power that a transceiver can transmit. The data rate is the amount of information that can be sent by the transceiver per unit time.

For duplex operations, transceivers must be used in matching pairs. They may use AM, FM, on-off keying (OOK), ASK, FSK, or PSK modes of operation.

2.3.17 Wireless Modems

Modems (MOdulator-DEModulator) are devices that allow digital devices to communicate over telephone lines. Wireless modems transmit data in electromagnetic waves or by means of optics. Wireless modems, also called radio modems, are RF transceivers that are connected to the serial ports (e.g., RS232, RS422) of digital devices. They transmit and/or receive signals from another matching radio modem. In many applications, a number of modems are connected to access points that enable wireless network connectivity.

Specifications for wireless modems are the form factor, modem speed, network type, and bus or interface type. In regards to the form factor, there are two basic types of modems: internal modems that are computer cards installed in a slot of the motherboard, and external modems that are modules that can be attached to communication ports of computers via cables. Modem speed is the maximum data transfer rate at which the modem can deliver data, expressed in bits per second. Network types for wireless modems include dial-up, Ethernet, Global System for Mobile Communications (GSM), integrated services digital network (ISDN), personal area network (PAN), general packet radio service (GPRS), and so on. The interface type modems include type II cards, type III cards, CardBus, FireWire (IEEE 1394), ISDN BRI U interface, serial ports (RS232, RS422, RS485), universal serial bus (USB), and power line communications (PLC) slot mount.

Modems are also identified by their radio link specifications, which include the frequency bands and the radio technique used. The frequency band on wireless modems can be 900 MHz, 2.4 GHz, 5 GHz, 23 GHz, VHF, and UHF. Choices for radio technique include direct sequence spread spectrum and frequency hopping spread spectrum, among others. Spread spectrum is a technique that is used to reduce the impact of localized frequency interferences by using more bandwidth than the system may need. There are two main spread spectrum modems: direct sequence modems and frequency hopping modems. The principle of direct sequence spreads the signal on a larger band by multiplexing it with a code to minimize localized interference and noise. Frequency hopping uses a technique where the signal uses a set of narrow frequency channels in sequence. The transmission frequency band is divided into a certain number of channels and periodically the system hops to a new channel, following a predetermined cyclic hopping pattern.

2.4 Analog Modulation and Multiplexing

Shifting or translation of a signal from one frequency to another frequency is achieved by the process of modulation. In modulation, a baseband signal (voice, video, etc.) is impressed on a carrier signal at a higher frequency, thus altering the frequency and amplitude characteristics of the carrier. Imposition of the baseband signal on the carrier signal is called modulation. Frequency translation from one to another is called continuous wave (CW) modulation. In RF communication systems, modulation permits the transmission of a message in a more efficient manner. In addition, multiplexing, such as FDM, can be used to allow several messages to be transmitted simultaneously by the same system.

Consider a carrier signal with frequency $_{c}$,

$$v_c(t) = V_c \sin(\omega_c t + \Phi_c), \tag{2.13}$$

where v_c is the instantaneous carrier voltage, V_c is the peak value of the voltage, ω_c is the angular frequency, and ϕ_c is the phase of the carrier.

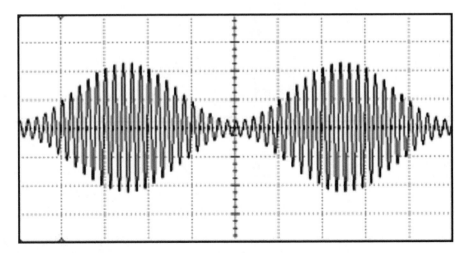

FIGURE 2.18
An AM waveform.

In the modulation process, one or more properties of the carrier signal—amplitude, frequency, or phase—can be used for modulation purposes. Consequently there are three major types of modulation techniques: AM, FM, and PM.

Modulation can also be digital. In digital modulation, a digital signal is used to modulate some properties of the carrier signal such as amplitude, frequency, phase, or a combination of these. Digital modulation is explained in Section 2.5.

2.4.1 Amplitude Modulation

In AM, the amplitude of the carrier is allowed to vary in proportion to the amplitude of the baseband signal. The amplitude modulated waveform is known as the modulation envelope, as illustrated in Figure 2.18. In this figure, the carrier frequency is modulated by a sinusoidal waveform.

When a baseband signal in the frequency range of f_l to f_h modulates the carrier signal having a fixed frequency, f_c, the resulting signal assumes a new frequency spectra as well as being subjected to alterations in bandwidth.

If the carrier signal is $V_c \sin(\omega_c t)$ and the modulating signal is $v_m(t)$, ignoring the phases, then the amplitude modulated signal can be expressed as

$$v(t) = [V_c + v_m(t)]\sin(\omega_c t). \qquad (2.14)$$

Assuming the modulating signal (e.g., the baseband signal) is $v_m(t) = V_m \sin(\omega_m t)$, then

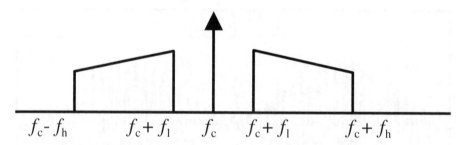

FIGURE 2.19
The frequency spectrum of an AM waveform.

$$v(t) = [V_c + V_m \sin(\omega_m t)]\sin(\omega_c t). \tag{2.15}$$

Expanding this equation and using trigonometric relationships results in

$$v(t) = V_c \sin(\omega_c t) + \frac{V_m}{2}\cos(\omega_c - \omega_m)t - \frac{V_m}{2}\cos(\omega_c + \omega_m)t . \tag{2.16}$$

The envelope has a maximum amplitude $(V_c + V_m)$ and a minimum amplitude $(V_c - V_m)$. It also contains three different frequencies: f_c, a lower side frequency $(f_c - f_m)$, and an upper side frequency $(f_c + f_m)$.

Although the carrier signal frequency is constant, the frequency of the modulating signal will have a frequency band between the upper frequency (f_h) and the lower frequency (f_l). Thus the modulated signal produces a lower sideband and an upper sideband in the frequency spectrum, as illustrated in Figure 2.19.

The bandwidth occupied by the modulated signal depends on the modulation scheme used. The required bandwidth for conventional AM is typically twice the highest modulating frequency. For SSB, the bandwidth is nominally equal to the baseband bandwidth.

The modulation index (k) of an AM signal is defined as the ratio of the peak baseband signal amplitude to the peak carrier signal amplitude. For a sinusoidal modulating signal, $v_m(t) = V_m\cos(2\pi f_m t)$, the modulation index is given by

$$k = V_m / V_c. \tag{2.17}$$

The modulation index is often expressed in percentage values, hence it is called the percentage modulation. The importance of the modulation index is that the relative power in the sidebands and the carrier signal depends on the percentage of modulation.

There are various types of AM, including double-sideband suppressed carrier, vestigial sideband (VSB), SSB, etc., which will not be explained here.

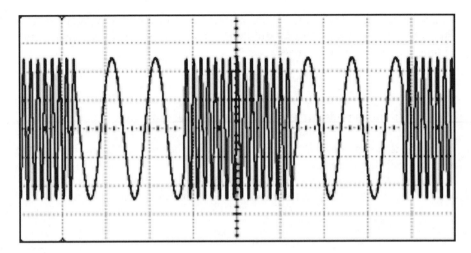

FIGURE 2.20
An FM waveform.

2.4.2 Frequency Modulation

In FM, the RF of the carrier is altered proportional to the amplitude of the modulating signal while keeping the amplitude of the carrier signal constant. Figure 2.20 illustrates a binary modulated FM signal.

The carrier wave has the highest frequency when the modulating waveform is at its maximum amplitude. Between the maximum and minimum amplitudes of the modulating signal, the frequency of the carrier changes to maximum and minimum above its resting frequency. This process is called frequency deviation.

If the carrier signal is $V_c\sin(\omega_c t)$ and the modulating signal is $V_m\sin(\omega_m t)$, ignoring the phases, the frequency of the modulated signal can be expressed as

$$f = f_c + f_\Delta V_m \sin(\omega_m t), \tag{2.18}$$

where $f\Delta$ is the frequency deviation constant. Hence a frequency modulated signal can be expressed as

$$v(t) = V_c\sin[2\pi(f_c + f\Delta V_m \sin 2\pi f_m t)t]. \tag{2.19}$$

This equation has an expression for the sine of a sine, which requires Bessel functions of the first order for its solution. By algebraic manipulation, this equation can be expressed in a more familiar form as

$$v(t) = V_2 \sin\left[2\pi f_c t + \frac{f_\Delta}{f_m} V_m \cos 2\pi f_m t \right]. \tag{2.20}$$

The frequency deviation (Δf) of a frequency modulated waveform is a measure of how much the modulated signal deviates from the carrier frequency, which can be expressed as

$$\Delta f = f_\Delta V_m. \tag{2.21}$$

In FM, there is no theoretical maximum value to the frequency deviation. However, for a given system, a maximum frequency deviation, called the rated system deviation (f_{max}) must be defined because of limitations in the operational bandwidths.

The modulation index (m_f) of an FM wave is the ratio of the frequency deviation of the carrier to the modulating frequency, that is,

$$m_f = \frac{\Delta f}{f_m}. \tag{2.22}$$

The modulation index determines the amplitudes of the frequency components of the modulated wave. It is also expressed in radians and is known as β.

From the above equations, it can be concluded that FM requires large bandwidths, hence it is inefficient in its use of spectrum space. However, FM demonstrates greatly improved performance in noisy environments. FM produces an infinite number of side frequencies spaced equally on both sides of the carrier frequency (f_c), as shown in Figure 2.21.

The frequency components in the FM spectrum eventually decrease to zero, but not in a straightforward way, fluctuating on either side of zero. However, for practical purposes, the bandwidth of the modulated signal can be approximated by $2(f_{m,max} + \Delta f)$, where Δf represents the peak frequency shift of the carrier from its rest frequency, which occurs when the baseband signal is at its maximum amplitude.

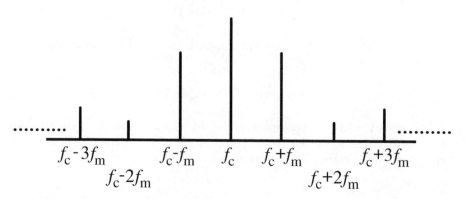

FIGURE 2.21
A typical frequency spectrum of an FM wave.

2.4.3 Phase Modulation

Analog PM is similar to FM. The instantaneous phase of the carrier signal is varied between the values of $(\phi + \Delta\phi)$ and $(\phi - \Delta\phi)$ at a rate proportional to the modulating frequency and by an amount equal to the amplitude of the modulating signal. The equation for PM can be expressed as

$$v(t) = V_c\sin(\omega_c t + \Phi_m\sin\omega_m t), \qquad (2.23)$$

where ϕ_m is the modulating index.

Despite the similarities between PM and FM, PM has a significant difference, that is, the maximum deviation in phase depends only on the amplitude of the modulating waveform and is independent of the frequency. Nevertheless, analog PM is rarely used in practice, except in some cases as an intermediate step in the production of FM.

In communication systems, many different signals can be transmitted on a single communication link by using suitable multiplexing techniques. There are two types of analog multiplexing: FDM and TDM, which are discussed in the next section.

2.4.4 Frequency and Time Division Multiplexing

Frequency division multiplexing band limits each channel to a maximum frequency, as illustrated in Figure 2.22a. Using FDM, many different signals (e.g., voices of multiple users) can be transmitted on a single communication link. Vacant frequency bands, called guard bands, are allocated between channels to prevent cross talk through overlapping of spectra of adjacent channels. This helps effective filtering and frequency selection on the receiver side. Each channel waveform is modulated onto a sinusoidal subcarrier to translate the channel to its allotted frequency position. In such systems, many subchannels are used in parallel in the main allocated channel, and thus they can transmit simultaneously over one physical circuit.

Time division multiplexing uses a single wideband channel for many input channels by assigning time slots for each channel. In TDM, at any instant in time, only one signal can be transmitted. As illustrated in Figure 2.22b, channel 1 occupies the entire channel during the first time slot, channel 2 occupies the entire channel during the second time slot, and so on.

A typical application is the analog wireless telemetry, which uses multiplexing techniques to transmit information acquired from instruments or sensors in remote locations. Telemeters represent an application of analog multiplexing techniques where there are many instruments involved. In telemetry, either FDM or TDM can be applied. With FDM, a number of separate frequency channels are used in parallel to transmit various channels of data. With TDM, the transmission of data alternates, allowing each instrument to send its own information through a single communication channel. Data channel 1 uses the single channel, then data channel 2 uses

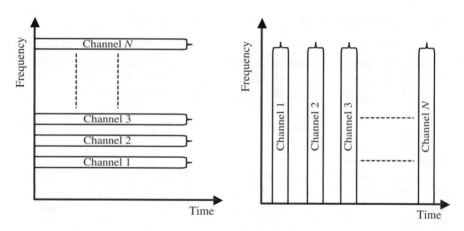

FIGURE 2.22
(a) Frequency division multiplexing. (b) Time division multiplexing.

the same frequency channel, then data channel 3, and so on, until all the channels have sent their information.

Many telemetry systems use both narrowband channels and relatively wideband channels. This mixture is achieved by a process known as subcommutation. Subcommutating consists of taking one of the wideband channels and subdividing it into several narrowband components.

2.5   Digital Modulation and Multiplexing

Because of recent progress in digital signal processing and associated technologies, modern communication systems have robust operations and better security in information flow. Security in information flow is enhanced by the application of cryptography, spread spectrum, and other techniques that could not be easily implemented in analog systems. Common to all communication systems, there are two major concerns to be addressed: use of minimum transmission bandwidth, and minimum error probability in the process of transmission. Digital communication systems can address both of these concerns and offer many additional advantages over analog counterparts.

Progress in digital modulation techniques is one of the major contributing factors in the wide applicability and use of digital communication systems. Digital modulation is a relatively new method that is quickly improving as the demand for digital communication increases. Digital modulation techniques can be divided in three major groups: amplitude shift keying (ASK), frequency shift keying (FSK), and phase shift keying (PSK), which are somewhat similar to AM, FM, and PM in analog modulation.

2.5.1 Amplitude Shift Keying

Amplitude shift keying is a simple form of AM that can be achieved by on-off keying of the carrier frequency, where on represents binary 1 and off represents binary 0. Although simple, the on-off keying method is not normally used in digital data transmission because of significant variations in the signal strength at the receiver.

An ASK data signal is generated by multiplying the data signal by a carrier, which mathematically can be expressed as

$$v_c(t) = h(t)\cos(\omega_c t), \tag{2.24}$$

where $h(t) = A$ or 0.

This multiplication effectively shifts the data spectrum to the center frequency equal to that of the carrier. The bandwidth of the modulated signal is twice the bandwidth of the original data signal.

Amplitude shift keying is not a desired form for transmitting data that has complex codes, such as Murray code. On-off keying presents significant noise problems at the receiver and introduces difficulties in determining transmitted binary 1s and 0s, particularly in noisy environments. In some simple cases, two-state or four-state AM is used in combination with pulse PM techniques to increase the number of bits per baud or bits per pulse that are transmitted.

2.5.2 Frequency Shift Keying

Frequency shift keying is the binary equivalent of analog FM. In this case, binary 0 is transmitted as a frequency f_0 and binary 1 is transmitted as a frequency f_1, while keeping the amplitude of the RF waveform constant, as illustrated in Figure 2.23. Thus the binary signal effectively modulates the carrier in a less complicated manner than a continuous analog baseband signal. In practice, the frequency shift is only a few kilohertz.

Frequency shift keying may be regarded as the sum of two ASK waveforms with different carrier frequencies. Hence the spectrum of the FSK wave is the sum of the spectra of the two ASK waveforms. The receiver is simply required to differentiate between the two frequencies f_0 and f_1 for successful reception. This is particularly important in mobile communication systems where there is an overriding need to minimize the transmitted bandwidth.

Standard FSK uses two separate carriers, f_0 and f_1, to transmit binary 0 and binary 1. In order to produce the smallest error probability, the carriers f_0 and f_1 must be orthogonal, that is, they must have a correlation coefficient of zero. In order to minimize the bandwidth of the transmitted signal, it is necessary to determine the minimum difference between frequencies f_0 and f_1, which produces orthogonal signals. This method is called minimum shift keying (MSK). If the number of cycles of f_0 in the bit period T is n_0, then f_1 will be orthogonal to f_0 if the number of cycles in the interval T is $n_1 = n_0 + 0.5$,

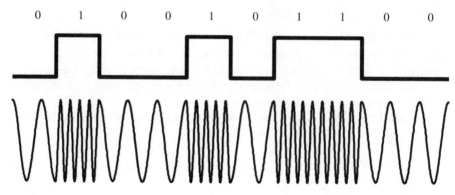

FIGURE 2.23
An FSK waveform.

alternatively $(f_1 - f_0) = 1/2T$. This indicates that MSK is basically a form of FSK with a minimum frequency difference between f_0 and f_1.

Frequency shift keying is the primary method of transmitting data over older radio communication links, since analog FM modulation can easily be replaced by FSK. Also, existing FM receivers capable of carrying analog data can be used. However, radio links specifically designed for data transmission largely employ the PSK technique because of its bandwidth requirements and power efficiency.

2.5.3 Phase Shift Modulation

Phase shift keying was developed for communication systems in early deep space exploration, but today, it is used extensively in military, industrial, and civilian communication. In this method, digital information is transmitted by shifting the phase of the carrier among several discrete values. The performance of binary PSK is similar to FSK. Nevertheless, there is one important difference between the two—the bandwidth required by the transmitted signal in PSK is much less than that in FSK because transmission can be done on only one frequency.

There are many variations of PSK: binary phase shift keying (BPSK), differential phase shift keying (DPSK), quaternary phase shift keying (QPSK), offset QPSK (OQPSK), differential QPSK (DQPSK), $\pi/4$ DQPSK, octonary phase shift keying (OPSK), etc. These modulation techniques form the backbone of modern communication systems.

2.5.4 Binary Phase Shift Keying

Binary phase shift keying is equivalent to analog PM, as it is another form of the two-state modulation technique. In BPSK, both the frequency and amplitude of the carrier remain constant, and only the phase of the signal

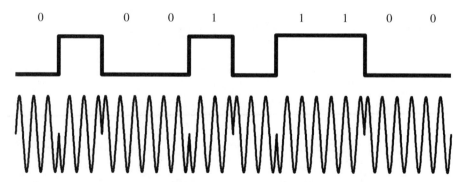

FIGURE 2.24
Amplitude shift keying.

is changed. The binary symbols are transmitted as a phase shift of zero or π radians, hence the two binary states are 180° apart in phase. A BPSK waveform can be expressed as

$$v_c(t) = h(t)\cos(\omega_c t), \tag{2.25}$$

where $h(t) = +A$ or $-A$.

A typical BPSK waveform and amplitude spectrum are illustrated in Figure 2.24.

The bandwidth of BPSK is similar to the bandwidth of ASK, however, because of the mean value of $h(t) = 0$, there is no component at the carrier frequency, f_c. Also, considerable difficulties can occur in generating the required carrier phase reference in some detectors, such as the coherent detectors. These difficulties are especially prominent in multipath environments, where multiples of the same signal are received spaced by time delays between the arrivals. This problem can be avoided by the use of DPSK.

In practical implementations, the reference signal is the signal supplied from an oscillator. A similar reference is used for both the transmitter and receiver.

2.5.5 Differential Phase Shift Keying

It is possible to send a BPSK signal as a DPSK signal. In DPSK, the information sent is determined by a change in the phases between the successive pulses. By using the phase of the carrier signal in the previous digit interval as the reference, the phase of the present digit interval can be determined. In order to make this possible, binary 0 is transmitted as the same phase as the previous digit and binary 1 is transmitted as a change in the phase. On the receiver side, the receiver constantly compares the phase of the current digit with the phase of the previous one. No change indicates binary 0. If

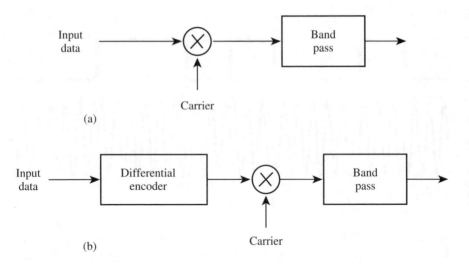

FIGURE 2.25
(a) Binary PSK and (b) DPSK modulators.

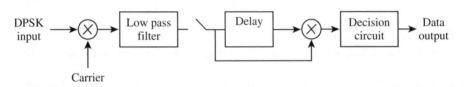

FIGURE 2.26
A DPSK demodulator.

there is a phase change, the digit is interpreted as a binary 1. Typical BPSK and DPSK encoders are illustrated in Figure 2.25.

In differential phase shift modulation, the phase reference at the receiver is the signal itself delayed by one pulse period, which can be achieved by analog or digital means. A typical DPSK decoder is shown in Figure 2.26. In this application, the output of the coherent detector is filtered and passed to a correlator, which is realized as a delay circuit and multiplier. If the phase is the same as the previous phase of the correlator, the output is positive and interpreted as binary 0. If the phase is the inverse of the previous phase, the output of the correlator is negative and is interpreted as binary 1. One advantage of this is that small phase errors in the carrier injected into the coherent detector do not affect correlator output.

2.5.6   Quaternary Phase Shift Keying

Figure 2.27 illustrates a typical four-state QPSK waveform. This type of modulation provides two or three bits of information for each pulse sent. The PSK waveform can be represented by the expression

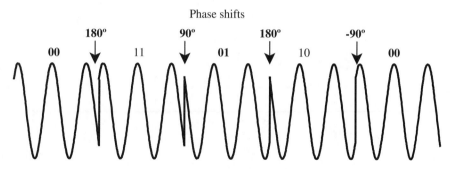

FIGURE 2.27
A typical four-state QPSK waveform.

TABLE 2.2

Phase shift in QPSK

Digital bits	Phase shift	In-phase amplitude	Quadrature amplitude
00	$\pi/4$	$+1/\sqrt{2}$	$+1/\sqrt{2}$
01	$3\pi/4$	$-1/\sqrt{2}$	$+1/\sqrt{2}$
11	$-3\pi/4$	$-1/\sqrt{2}$	$-1/\sqrt{2}$
10	$-\pi/4$	$+1/\sqrt{2}$	$-1/\sqrt{2}$

$$v_c(t) = V_c\cos(\omega_c t + \phi_k). \tag{2.26}$$

For BPSK, ϕ_k takes on two possible values separated by 180°. In four-state modulation—QPSK—ϕ_k takes on two possible values for phases separated by 90°. QPSK is equivalent to PM, but with four possible levels, as shown in Table 2.2.

Equation 2.26 can be expanded as

$$v_c(t) = V_i\cos(2\pi f_c t) + V_q\sin(2\pi f_c t), \tag{2.27}$$

where $V_i = V_c\cos(\phi_k)$ represents the amplitude of the in-phase component and $V_q = -V_c\sin(\phi_k)$ represents the amplitude of the quadrature component. The modulated carrier waveform, $v_c(t)$, has a constant amplitude only varying in phase, hence it is called constant envelope modulation.

In QPSK, the input data stream is grouped into dibits. Each group of digits produces a unique phase shift of the carrier, separated by an interval of $\pi/2$. Table 2.2 lists the possible dibits and their values in in-phase and quadrature components, V_i and V_q, respectively, and the associated phase shift. The resulting QPSK can be represented on a diagram as shown in Figure 2.28. A typical QPSK modulator, based on the dibit principle, is shown in Figure 2.29.

On the receiving side, the data signal is recovered from the QPSK waveform by using two coherent detectors that supplied oscillators generating in-phase

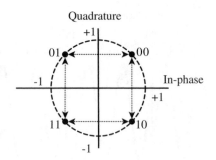

FIGURE 2.28
An I/Q diagram for QPSK.

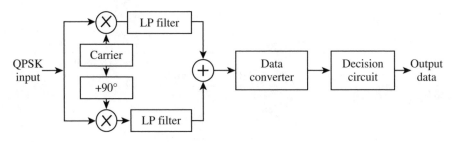

FIGURE 2.29
A typical QPSK modulator using dibits.

and quadrature components of the carrier wave. This process produces an in-phase data waveform and a quadrature data waveform separately. A drawback of this method is that a single error made in one of the digits in the detection process affects two digits in the reproduced data waveform.

2.5.7 Offset QPSK

In many communication systems with frequencies in the microwave region (e.g., mobile systems), the amplifiers in the circuits tend to have nonlinear characteristics. Nonlinear amplifiers are suitable for constant envelope modulation systems, thus QPSK is a suitable modulation scheme for such applications. However, when pulse shaping is applied to limit the bandwidths of the transmitted waveforms, it produces amplitude variations in the QPSK carrier envelope because of the nonlinearity. In many applications, these amplitude variations are largely suppressed, resulting in QPSK signals that look almost like a constant envelope. Unfortunately this can negate the advantages of the pulse-shaping operations in reducing the bandwidths of the transmitted waveforms.

A special form of QPSK, known as OQPSK, addresses this problem, as illustrated in Figure 2.30. In OQPSK, the data for the quadrature channel is delayed by a pulse duration T. This results in the maximum phase duration

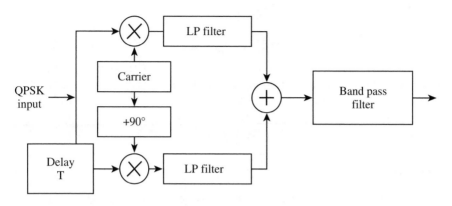

FIGURE 2.30
An offset QPSK modulator.

of the carrier signal, thus restricting it to 50% of the phase transition of a standard QPSK. As a result, a much smaller envelope variation takes place after band-pass filtering. Hence the phase transition, which passes through the origin, is removed by the delay of T, thus restricting the maximum phase shift to $\pi/2$.

2.5.8 Differential QPSK

Data recovery from the QPSK waveform requires two quadrature coherent detectors supplied with the appropriate reference carrier signals. Achieving an appropriate reference signal may be difficult in environments where multipaths exist. The problem may be addressed by the use of DQPSK. A typical DPSK modulation process reproduces the in-phase and quadrature data waveforms, each of which has a pulse duration of $2T$. The original input data with a pulse duration T may then be derived. A DQPSK demodulator is illustrated in Figure 2.31.

$\pi/4$ DQPSK is an improved form of DQPSK. In this case, each dibit combination produces a specific phase transition multiple of $\pi/4$. The phase transition is independent of the current carrier phase and has a maximum

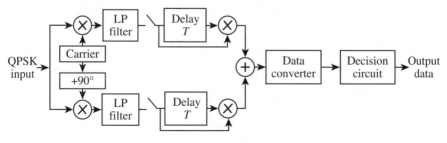

FIGURE 2.31
A DQPSK demodulator.

value of $3\pi/4$, which is less than the maximum phase shift produced in OPSK. Although there are eight separate carrier phases, there are only four possible phase transitions: 00, 01, 10, and 11.

At the receiver, a particular dibit is produced by comparing the current carrier phase with the previous carrier phase and measuring the phase differences. This simple operation requires the presence of both in-phase and quadrature phase components in order to differentiate between phase transitions $\pi/4$ and $-\pi/4$. For this operation, slightly more complex decoders are used.

2.5.9 Octonary Phase Shift Keying

In the case of eight-state, also known as octonary phase shift keying (OPSK), phase shift steps are separated by 45°. In this case, three bits of information can be sent per pulse.

In 16-phase state keying (16-PSK), each phase step is separated by 22.5°, thus allowing the transfer of four bits of information per pulse. An advanced version of 16-PSK is 16-amplitude phase keying, which is often used in high-capacity TDM systems.

2.5.10 Digital Multiplexing

As in the case of analog systems, digital multiplexing is based on FDM or TDM techniques. Although, in digital transmission, both FDM and TDM techniques are used, FDM is preferred for slow data flow between terminals spread over wide geographical areas. TDM is generally applied in high data rate terminals clustered in one location or in situations where long-distance digital signal transmission is required. Table 2.3 compares FDM and TDM use in signal transmission.

2.6 Frequency Spreading and Multiple Access Techniques

Since in RF systems the bandwidth is a limited resource, one of the primary design objectives of modulation techniques is to minimize the required trans-

TABLE 2.3

Comparison of FDM and TDM

Function	FDM	TDM
Signal type	Analog, channels carry original waveform	Digital, channels carry waveform samples
Speed	Slow, depends on bandwidth availability	High, depends on channel and modem
Transmission	Parallel	Serial
Network type	Wide spread	Clustered

mission bandwidth. Spread spectrum techniques require transmission bandwidths several orders of magnitude greater than the minimum required signal bandwidth. Clearly this system is very bandwidth inefficient for a single user, but it offers advantages in allowing many users to use the same bandwidth without significantly interfering with one another. In a multiuser system, spread spectrum becomes very bandwidth efficient. Spread spectrum also solves many problems generated in multiple access interference (MAI) environments.

Apart from occupying a very large bandwidth, spread spectrum signals are pseudorandom signals that have noise-like properties when compared to digital information data. The spreading of the waveform is controlled by a pseudonoise sequence or pseudonoise code, which is essentially a binary sequence that appears random but can be reproduced in a deterministic manner by the intended receiver. Spread spectrum signals are demodulated at the receiver side through a cross-correlation process with the locally generated version of the pseudorandom carrier. Cross correlation with the pseudonoise sequence despreads the received spread spectrum signal and restores the modulated message in the same narrow band as the original data.

A pseudonoise (PN) is essentially a binary sequence with an autocorrelation process that resembles the autocorrelation of a random binary sequence. Its autocorrelation also roughly resembles the autocorrelation of band-limited white noise. Although it is deterministic, a pseudonoise sequence, as a random binary sequence, has many characteristics, such as having a nearly equal number of 0s and 1s, a very low correlation between shifted versions of the sequence, and a very low correlation between any two sequences.

Spread spectrum has an inherent interference rejection capability. Each user is assigned a unique pseudonoise code that is approximately orthogonal to the codes of the other users. The receiver is able to separate each user based on the information on their allocated codes, even though they occupy the same spectrum at all times. Resistance to multipath fading is another useful property of the spread spectrum technique. Since spread spectrum signals have uniform energy over a very large bandwidth at any given time, only a small portion of the spectrum undergoes fading. In time domain, the multipath resistance properties are due to the fact that the delayed versions of the transmitted pseudonoise signal will have poor correlation with the original sequence, and thus appear as another uncorrelated user that is ignored by the receiver.

Multipath signals arrive at the receiver in a delayed fashion, which can be used to improve the performance of the system. This can be done by using rake receivers, which anticipate the multipath delays of the transmitted spread spectrum signal and combine the information obtained from several resolvable multipath components to form a stronger version of the signal. A rake receiver consists of a bank of correlators, each of which correlates to a particular multipath component of the desired signal. The correlator out-

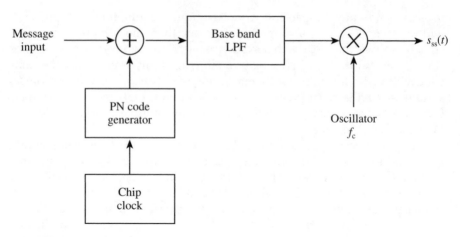

FIGURE 2.32
A block diagram of a DSSS transmitter with binary PM.

puts may be weighted according to their relative strengths and summed to obtain a final estimate of the signal.

There are two main versions of spread spectrum techniques: direct sequence spread spectrum (DSSS) and the frequency hopped spread spectrum (FHSS).

2.6.1 Direct Sequence Spread Spectrum

A DSSS system spreads the baseband data by directly multiplying data pulses with a pseudonoise sequence. The PN sequence is produced by an internal code generator. A single pulse or symbol of the pseudonoise waveform is called a chip. Synchronized data symbols, information bits, or binary channel code symbols are added to the chip before being phase modulated. Figure 2.32 shows a block diagram of a DSSS system using binary PM at the transmitter.

The receiver demodulates the transmitted signal in a number of different forms, such as by using coherent PSK demodulators or differentially coherent PSK demodulators. For a single user, the received spread spectrum signal can be represented as

$$s_{ss}(t) = \sqrt{\frac{2E_s}{T_s}}\, m(t)\, p(t)\cos\left(2\pi f_c t + \theta\right), \tag{2.28}$$

where $m(t)$ is the data sequence, $p(t)$ is the pseudonoise spreading sequence, f_c is the carrier frequency, and θ is the carrier phase angle at $t = 0$.

Figure 2.33 illustrates a DSSS receiver. Assuming that clock synchronization has been achieved by the receiver, the received signal passes through

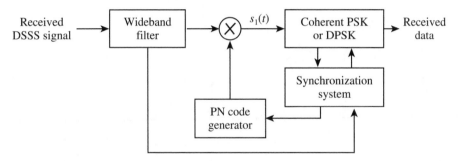

FIGURE 2.33
A block diagram of a DSSS receiver with binary PM.

the wideband filter and is multiplied by a local replica of the pseudonoise code sequence, $p(t)$. If $p(t) = \pm1$, then $p^2(t) = 1$ and this multiplication yields the despread signal $s(t)$ at the input of the modulator. This can be expressed as

$$s_1(t) = \sqrt{\frac{2E_s}{T_s}} m(t) \cos\left(2\pi f_c t + \theta\right). \tag{2.29}$$

Since $s_1(t)$ has the form of a BPSK signal, the corresponding demodulation can easily extract $m(t)$.

Figure 2.34a shows the received spectra of the desired spread spectrum signal and the interference at the output of the receiver wideband filter. Multiplication by the spreading waveform produces the spectra (see Figure 2.34b) at the input of the demodulator. The signal bandwidth is reduced to instantaneous bandwidth B, while interference energy is spread over a total RF bandwidth exceeding B_{ss}. The filtering action of the demodulator removes most of the interference spectrum. This process is very important, as the interference energy is eliminated by spreading, thus minimally affecting the desired receiver signal. An approximate measure of the interference rejection capability is equal to the processing gain, which is defined as PG = $\dfrac{B_{ss}}{B}$.

$$PG = \frac{B_{ss}}{B}. \tag{2.30}$$

The greater the processing gain of the system, the better its ability to suppress in-band interference.

A DSSS system with K multiple access is shown in Figure 2.35. Assume each user has a pseudonoise sequence with N chips per message symbol

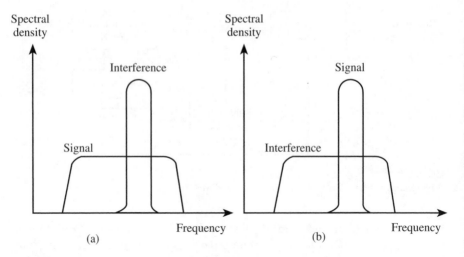

FIGURE 2.34
Spectra of a received signal: (a) wideband filter output and (b) correlator output.

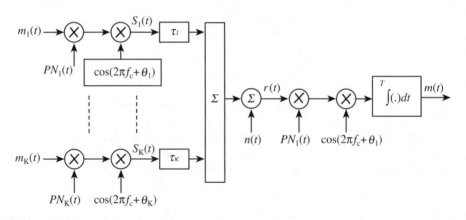

FIGURE 2.35
A simplified diagram of DSSS with multiple users.

period, T_c, such that $NT_c = T$. The transmitted signal of the kth user can be expressed by

$$S_k(t) = \sqrt{\frac{2E_s}{T_s}} m_k(t) p_k(t) \cos\left(2\pi f_c t + \theta_k\right),$$ (2.31)

where $p_k(t)$ is the pseudonoise code sequence of the kth user and $m_k(t)$ is the data sequence of the kth user.

Successful reception is accomplished by correlating the received signal with an appropriate signature sequence to produce a decision variable. For user 1, the decision variable can be expressed as

$$Z_i = \int_{(i-1)T+\tau_1}^{iT+\tau_1} r(t)p_1(t-\tau_1)\cos\left[2\pi f_c(t-\tau_1+\theta_1)\right]dt . \tag{2.32}$$

From the decision variable, multiple access interference, I_k, from user k and the Gaussian random variable (E) representing noise with mean zero and variance can be calculated as

$$I_k = \int_0^T S_k(t-\tau_k)p_1(t)\cos(2\pi f_c t)dt \tag{2.33}$$

and

$$E = \frac{N_0 T}{4} . \tag{2.34}$$

Assuming that I_k is composed of cumulative effects of N random chips from the kth interferer over the integration period T of one bit, the central limit theorem applies and some of these effects tend toward the Gaussian distribution. Since there is $K-1$ users that serve as identically distributed interferers, the total multiple access interference may be approximated by a Gaussian random variable. The Gaussian approximation yields a convenient expression for the average probability of bit error that can be expressed as

$$P_e = Q\left(\frac{1}{\sqrt{\dfrac{K-1}{3N}+\dfrac{N_0}{2E_b}}}\right) . \tag{2.35}$$

For a single user, $K = 1$, for the interference limited case, thermal noise is not a factor and N_0/E_b tends to zero, hence this expression reduces to

$$P_e = Q\sqrt{\frac{3N}{K-1}} . \tag{2.36}$$

This is the irreducible error floor due to multiple access interference, and assumes that all interferers provide equal power, the same as the desired user, at the receiver. In practice, the near-far presents difficulty for DSSS

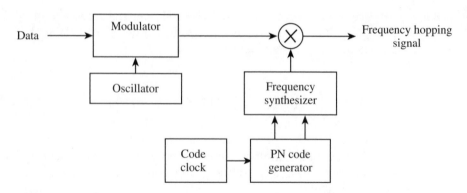

FIGURE 2.36
A block diagram of a frequency hopping transmitter.

systems. For example, one close-in user may dominate the received signal at the base station, making the Gaussian assumption inaccurate. For a large number of users, the bit error rate is limited more by multiple access interference than by thermal noise.

2.6.2 Frequency Hopped Spread Spectrum

Frequency hopping involves periodic changes in transmission frequency within an allocated frequency band. A frequency hopping signal may be regarded as a sequence of modulated data bursts varying in time at pseudorandom carrier frequencies. Hopping occurs over a frequency band that includes a number of set channels, each operating at an allocated frequency. The set of possible carrier frequencies is called a hop set. Each channel is defined as a spectral region with a central frequency in the hop set and a bandwidth large enough to include most of the power in a narrow modulation burst (e.g., FSK) within the corresponding carrier frequency. The bandwidth of a channel used in the hop set is called the instantaneous bandwidth. The bandwidth of the spectrum over which the hopping occurs is called the total hopping bandwidth. Data are sent by hopping the transmitter carrier to seemingly random channels that are known only by the intended receiver. On each channel, small bursts of data are sent using conventional narrowband modulations before the transmitter hops again.

When only a single carrier frequency is used in each hop, digital data modulation is called single-channel modulation. Figure 2.36 shows a single-channel FHSS system. The time duration between hops is called the hop duration or hopping period. The total hopping bandwidth, B_{ss}, and instantaneous bandwidths, B, yield the processing gain, B_{ss}/B, for the system.

On the receiver side in Figure 2.37, after frequency hopping has been removed from the received signal, the resulting signal is said to be dehopped. Before demodulation, the dehopped signal is applied to a conventional receiver. The frequency pattern produced by the receiver synthesizer is syn-

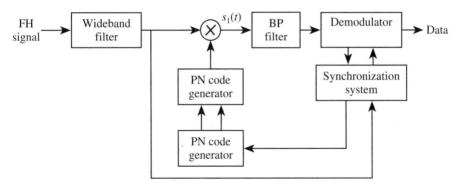

FIGURE 2.37
A block diagram of a frequency hopping receiver.

chronized with the frequency pattern of the received signal, the mixer output is a dehopped signal at a fixed difference frequency. Whenever an undesired signal occupies a particular hopping channel in the input signal, the noise and interference in that channel are translated in frequency so that they enter the demodulator. Thus, in this process, it is possible to have collision in a frequency hopping system where an undesired user transmits in the same channel at the same time as the desired user.

If binary FSK is used, the pair of possible instantaneous frequencies changes with each hop. The frequency channel occupied by a transmitted signal is called the transmission channel. The channel that would be occupied if an alternative signal were transmitted is called the complementary channel. The frequency hop rate of an FHSS system is determined by the frequency agility of the receiver synthesizer, the type of information being transmitted, the amount or redundancy used to code against collision, and the distance to the nearest potential interferer.

In FHSS systems, several users can independently hop their carrier frequencies while using binary FSK modulation. If two users are simultaneously utilizing the same frequency band, the probability of error can be written as

$$P_e = \frac{1}{2}\exp\left(-\frac{E_b}{2N_0}\right). \tag{2.37}$$

Error analysis can be extended for various scenarios such as synchronized frequency hopping (called slotted frequency hopping) and asynchronized operations by users to derive the irreducible error rates for the multiple access interferences and other detailed analysis.

Frequency hopped spread spectrum has an advantage over DSSS in that it is not as susceptible to the near-far problem. Because signals are generally not utilizing the same frequency simultaneously, the relative power levels of signals are not as critical as in DSSS. The near-far problem is not totally avoided, however, since there will be some interference caused by stronger signals bleeding into weaker signals due to imperfect filtering of adjacent channels. To combat the occasional hits, error correction coding is required on all transmissions. By applying appropriate (e.g., Reed-Solomon) error correcting codes, performance can be increased significantly.

2.6.3 Multiple Access Techniques

Multiple access schemes are used to allow many users to simultaneously share a finite amount of radio spectrum. Sharing is required to achieve high-capacity usage of the available bandwidth. In wireless communication systems, for example, it is desirable to allow users to send information to a station while receiving from the station at the same time. This can be achieved by duplexing, which can be done using frequency or time domain techniques.

Multiple access techniques can be grouped into narrowband and wideband systems, which depends on how the available bandwidth is allocated to the user. Both narrowband and wideband systems are supported by duplexing techniques such as frequency division duplexing (FDD) and time division duplexing (TDD).

Narrowband system is a term used to relate the bandwidth of a single channel to the expected coherence bandwidth of the channel. In a narrowband multiple access system, the available radio spectrum is divided into a large number of narrowband channels. The channels are usually operated using FDD. To minimize interference between the forward and reverse links on each channel, the frequency separation is made as large as possible within the frequency spectrum, while still allowing inexpensive duplexers and a common transceiver antenna in each unit.

Wideband systems have a transmission bandwidth of a single channel that is much larger than the coherence bandwidth of the channel. Thus multipath fading does not greatly vary the received signal power within a wideband channel, and the frequency selective fades occur in only a small fraction of the single bandwidth at any point in time. In wideband multiple access systems, a large number of transmitters are allowed to transmit on the same channel.

Frequency division duplexing provides two distinct bands of frequency: one for the transmission channel of the transmitter and one for the reception channel of the receiver (see Figure 2.38a). In FDD, separate transmit and receive antennas are used to accommodate the two separate channels. However, in some cases, a single antenna is used that is backed up by duplexers for simultaneous transmission and reception. To facilitate FDD, it is necessary to separate the transmit and receive frequencies by about 5% of the nominal RF so that the duplexer can provide sufficient isolation.

In FDD, a pair of simplex channels with a fixed, known frequency separation is used to define a specific radio channel in the system, called the

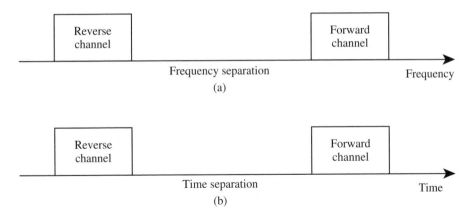

FIGURE 2.38
FDD and TDD duplexers.

forward channel for transmission and the reverse channel for reception. FDD is used largely in analog communication systems. In FDD, any duplex channel actually consists of two simplex channels and the frequency separation between the forward and reverse channels is constant.

Time division duplexing uses time instead of frequency to provide both a forward and reverse link. In TDD, multiple users share a single radio channel by taking turns in the time domain (see Figure 2.38b). As the forward and reverse links share a single radio channel in time, a portion of the time is used to transmit and the remaining time is used to receive within the allocated time slots. Each duplex channel has both a forward time slot and a reverse time slot to facilitate bidirectional communication. If the data transmission rate in the channel is much greater than the end user's data rate, it is possible to store information bursts and provide the appearance of full duplex operation even though there are not two simultaneous radio transmissions at any instant. TDD is only possible with digital transmission systems and digital modulation, and it is very sensitive to timing. TDD is used in indoor and small area wireless networks where physical coverage distances are relatively small.

There are several trade-offs between TDD and FDD. TDD enables each transceiver to operate as either a transmitter or a receiver on the same frequency. This eliminates the need for separate forward and reverse frequency bands. There is a time latency created by TDD due to the fact that communication is not full duplex in the truest sense, and this latency can create inherent sensitivity for propagation delays for individual users. Because of the rigid timing required for time slotting, TDD is generally limited to short-range access. TDD is effective for fixed wireless access when all users are stationary so that propagation delays do not vary in time among the users. FDD, on the other hand, is geared toward radio communication systems that allocate individual RFs for each user. Because each transceiver simultaneously transmits and receives RF signals, the frequency allocation

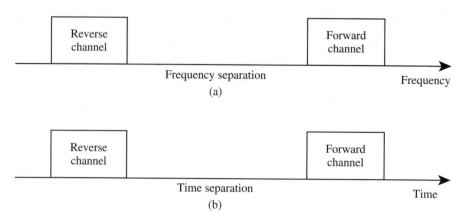

FIGURE 2.39
Channel and frequency assignment in FDMA.

in the forward and reverse channels must carefully be coordinated within the system as well as with out-of-band users that occupy the same spectrum. Furthermore, frequency separation must be supported by inexpensive RF technology.

In recent years, new and more effective multiple access techniques have been introduced. Consequently there are many different types that are currently operational, including frequency division multiple access (FDMA), time division multiple access (TDMA), spread spectrum multiple access (SSMA), code division multiple access (CDMA), frequency hopped multiple access (FHMA), hybrid spread spectrum multiple access, space division multiple access (SDMA), and carrier sense multiple access (CSMA).

2.6.4 Frequency Division Multiple Access

Frequency division multiple access assigns individual channels to individual users (see Figure 2.39). These channels are assigned on demand to users who request service. Upon request, users are assigned channels as a pair of frequencies for forward channel and reverse channel. Once allocated, no other user can share the same channel during the transmission.

Frequency division multiple access requires tight RF filtering to minimize adjacent channel interference. Also, if a channel is not used, it is said to be idle and cannot be used by others to increase or share the capacity. In order to increase capacity, techniques such as FDD can be used. Then the system is an FDMA/FDD system.

2.6.5 Time Division Multiple Access

In TDMA, each user occupies a cyclically repeating time slot, as illustrated in Figure 2.40. It can be thought of as having frames that comprise *N* reoc-

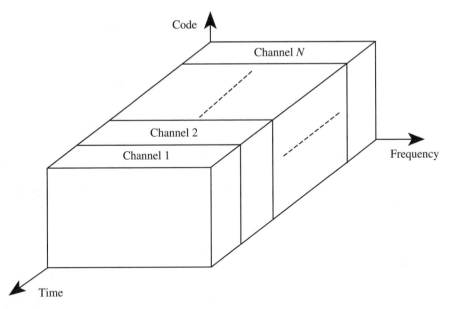

FIGURE 2.40
Channel and time slot assignment in TDMA.

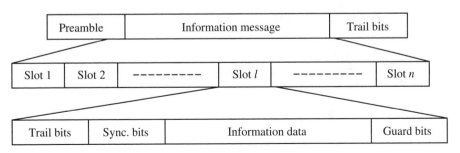

FIGURE 2.41
TDMA frame structure.

curring time slots. TDMA systems transmit data in a buffer-and-burst method, so the transmission for any one user is noncontinuous. Unlike FDMA systems, which accommodate analog FM, in TDMA, digital data and digital modulation must be used.

The transmission sequences from various users are interlaced in a repeating frame structure, as is illustrated in Figure 2.41. Each frame is made up of a number of sections: a preamble, an information message, and tail bits. The preamble contains the address and synchronization information that both the base station and other units use to identify each other. Guard times are utilized to allow synchronization of the receivers between different slots and frames. Different TDMA wireless standards have different frame structures.

In TDMA systems, the number of time slots per frame depends on the modulation techniques and the bandwidth available. Adaptive equalization is usually necessary because the transmission rates are generally higher compared to FDMA channels. Also, high synchronization overhead is required because of burst transmissions. TDMA transmissions are slotted, and this requires the receiver to be synchronized for each data burst. In addition, guard slots are necessary to separate users. Guard times should be minimized, but shortening guard times can cause interference in adjacent channels.

The efficiency of a TDMA system is a measure of the percentage of transmitted data that contains information as opposed to providing overhead for the access scheme. The frame efficiency, $_f$, is the percentage of bits per frame that contain transmitted data and can be expressed as

$$N_f = \left(1 - \frac{b_{oh}}{b_T}\right) \times 100 \, , \qquad (2.38)$$

where b_{oh} is the number of overhead bits per frame and b_T is the total number of bits per frame.

The number of overhead bits per frame can be expressed as

$$b_{oh} = N_r b_r + N_t b_p + N_t b_g + N_r b_g, \qquad (2.39)$$

where N_r is the number of reference bursts per frame, N_t is the number of traffic bursts per frame, b_r is the number of overhead bits per reference burst, b_p is the number of overhead bits per preamble in each slot, and b_g is the number of equivalent bits in each guard time interval.

The total number of bits per frame, b_T, is

$$b_T = T_f R, \qquad (2.40)$$

where T_f is the frame duration and R is the channel bit rate.

Time division multiple access is applied in narrowband systems as well as wideband systems. Narrowband TDMA allows users to share the same radio channel, but allocates a unique time slot to each user in a cyclic fashion on the channel. For narrowband TDMA systems, there generally are a large number of radio channels allocated using either FDD or TDD, and channels are shared using TDMA. Such systems are called TDMA/FDD or TDMA/TDD access systems.

In TDMA/TDD, half of the time slots in the frame information are used for forward link channels and half are used for reverse link channels. In TDMA/FDD systems, an identical or similar frame structure is used solely for either forward or reverse transmission, and the carrier frequencies are different for the forward and reverse links. In general, TDMA/FDD systems intentionally induce several time slots of delay between the forward and reverse time slots for a particular user so that duplexers are not required.

In wideband systems, TDMA allocates time slots to the many transmitters on the same channel and allows only one transmitter to access the channel at any point in time. In comparison, the spread spectrum of CDMA allows all the transmitters to access the channel at the same time. TDMA and CDMA systems use either FDD or TDD multiplexing techniques.

2.6.6 Spread Spectrum Multiple Access

Spread spectrum multiple access uses signals that have a transmission bandwidth that is several orders of magnitude greater than the minimum required RF bandwidth. A pseudonoise sequence converts the narrowband signal to a wideband noise-like signal before transmission. SSMA also provides immunity to multipath interference and robust multiple access stability. Since many users can share the same system, SSMA is bandwidth efficient and is used by many wireless communication designers.

There are two main types of SSMA techniques: FHMA and SDMA, also known as CDMA.

2.6.7 Code Division Multiple Access

In CDMA systems, the narrowband message signal is multiplied by a very large bandwidth signal called the spreading signal. The spreading signal is a pseudonoise code sequence that has a chip rate greater by orders of magnitude than the data rate of the message. As seen in Figure 2.42, all users in a CDMA system use the same carrier frequency and may transmit simultaneously. Each user has its own pseudorandom codeword which is approximately orthogonal to all other code words. The receiver performs a time correlation operation to detect only the specific desired codeword. All other code words appear as noise due to decorrelation. For the detection of a message signal, the receiver needs to know the codeword used by the transmitter. Each user operates independently, with no knowledge of the other users.

In CDMA, the power of multiple users at a receiver determines the noise floor after decorrelation. If the power of each user within a cell is not controlled so that they do not appear equal at the base station receiver, then the near-far problem occurs.

The near-far problem appears when many units share the same channel. In general, the strongest received signal captures the demodulation at the base station. In CDMA, stronger received signal levels raise the noise floor at the base station demodulators for weaker signals, thereby decreasing the probability that weaker signals will be received. To combat the near-far problem, a power control scheme is used in most CDMA implementations. Power control is provided by each base station and ensures that

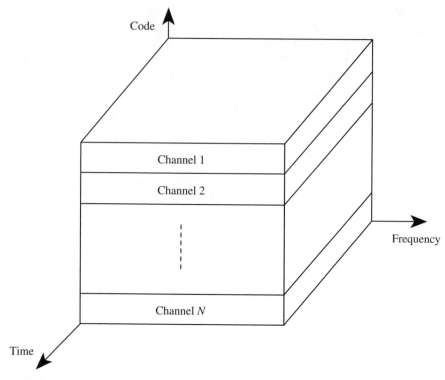

FIGURE 2.42
Spread spectrum multiple access CDMA.

each unit provides the same signal level to the receiver in a coverage area. Power control is implemented by the rapidly sampled signal indicator (RSSI) levels of each unit and sensing a power channel over the forward radio link.

In CDMA systems, multipath fading may be substantially reduced because the signal is spread over a wide spectrum. The channel data rates are high, consequently the symbol-time duration is low, much less than the channel delay spread.

2.6.8 Frequency Hopped Multiple Access

Frequency hopped multiple access is a digital multiple access system in which the carrier frequencies of the individual users are varied in a pseudorandom fashion within a wideband channel. Figure 2.42 also illustrates the way that FHMA allows multiple users to simultaneously occupy the same spectrum at the same time, where each user dwells at a specific narrowband channel at a particular point in time, based on the particular pseudonoise code of the user. The digital data of each user are broken into uniform-size bursts which are transmitted on different channels within the allocated spectrum band. The instantaneous bandwidth of any one transmission burst is much smaller than the total spread bandwidth. The pseudorandom changes of the channel frequencies of the user randomize the occupancy of the specific channel at any given time, thereby allowing for multiple access over a wide range of frequencies. On the receiver side, a locally generated pseudonoise code is used to synchronize the receiver's instantaneous frequency with that of the transmitter. At any given point in time, a frequency hopped signal only occupies a single, relatively narrow channel, since narrowband FM or FSK is used.

The difference between FHMA and traditional FDMA is that the frequency hopped signal changes channels at rapid intervals. If the rate of change of the carrier frequency is greater than the symbol rate, then the system is referred to as a fast frequency hopping system. If the channel changes at a rate less than or equal to the symbol rate, it is called slow frequency hopping. A fast frequency hopper may thus be thought of as an FDMA system that employs frequency diversity. FHMA systems often employ energy efficient constant envelope modulation.

Frequency hopped systems provide a level of security, especially when a large number of channels are used, since an unintended or interception receiver does not know the pseudorandom sequence of frequency slots must retune rapidly to search for the signal it wishes to intercept. In addition, the frequency hopped signal is somewhat immune to fading, since error control coding and interleaving can be used to protect the frequency hopped signal against deep fades, which may occasionally occur during the hopping sequence. Error control coding and interleaving can also be combined to guard against erasures that can occur when two or more users transmit on the same channel at the same time. Bluetooth and HomeRF wireless technologies have adopted FHMA for power efficiency and low-cost implementation.

2.6.9 Hybrid Spread Spectrum Techniques

In addition to frequency hopped and direct sequence spread spectrum multiple access techniques, there are certain hybrid combinations of multiple access methods that provide advantages.

The hybrid FDMA/CDMA (FSDMA) technique is used as an alternative to direct sequence CDMA (DS-CDMA). Figure 2.43 shows this hybrid scheme. The available bandwidth spectrum is divided into a number of subspectra. Each of the subchannels becomes a narrowband CDMA system having a processing gain lower than the original CDMA. This hybrid system has the advantage that different users can be allocated different subspectra depending on their requirements.

The hybrid DS-FHMA technique consists of a direct sequence modulated signal whose central frequency is made to hop periodically in a pseudorandom fashion. Figure 2.44 shows the frequency spectrum of such signals.

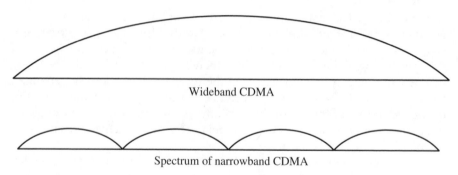

FIGURE 2.43
Spectra of wideband CDMA and hybrid narrowband CDMA.

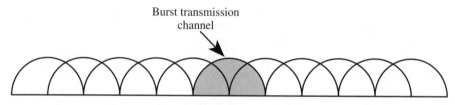

FIGURE 2.44
Spectrum of a hybrid DS-FDMA system.

Direct sequence frequency hopped systems have an advantage in that they avoid the near-far effect. However, frequency hopped CDMA systems are not easy to synchronize.

Time division CDMA (TCDMA) or TDMA/CDMA systems use different spreading codes for different cells. Within each cell, only one user per cell is allotted to a particular time slot. Thus, at any time, only one CDMA user can transmit in each cell. It avoids the near-far problem, since only one user can transmit within a cell.

Time division frequency hopping (TDFH) is a multiple access technique where the spectrum hops to a new frequency at the start of a new TDMA frame. It is used when severe multipath problems or severe cochannel interference occurs.

2.6.10 Space Division Multiple Access

Space division multiple access controls the radiated energy for each user by using spot beams (Figure 2.45). Different areas covered by different antenna beams may be served by the same frequency as in TDMA or CDMA, or by different frequencies as in FDMA. Extensive studies have concentrated on adaptive antennas to simultaneously steer energy in the direction of users. Adaptive antennas used at the base station promise to mitigate some of the

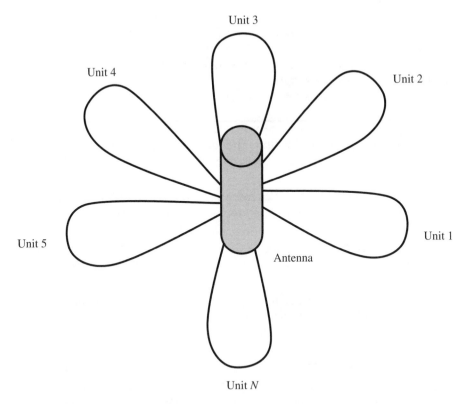

FIGURE 2.45
An SDMA antenna.

problems on the reverse link. In the limiting case of infinitesimal beam width and infinitely fast tracking ability, adaptive antennas implement optimal SDMA, thereby providing a unique channel that is free from the interference of all other users in the cell. With SDMA, all users within the system are able to communicate at the same time using the same channel. Perfect adaptive antennas do not exist at this time, but reasonable-size arrays with moderate directivities do exist.

2.6.11 Carrier Sense Multiple Access

Carrier sense multiple access protocols are based on the fact that each terminal on the network is able to monitor the status of each channel before transmitting information. If the channel is idle, then the user is allowed to transmit a packet based on a particular algorithm that is common to all transmitters on the network.

In CSMA protocols, detection delay and propagation delay are two important parameters. Detection delay is a function of the receiver hardware and is the time required for a terminal to sense whether or not the channel is idle. Propagation delay is a relative measure of how fast the signal travels from a base station to a mobile terminal. With small detection times, a terminal detects a free channel quite rapidly, and small propagation means that a packet is transmitted through the channel in a small time interval relative to the packet duration.

There are several variations of the CSMA strategy:

- 1-persistent CSMA—the terminal listens to the channel and waits for transmission until it finds the idle channel. As soon as the idle channel is identified, the transmitter transmits with probability of one.

- Nonpersistent CSMA—after receiving a not acknowledged (NACK) signal, the terminal waits a random time before retransmission. This is a popular technique in wireless local area network applications.

- P-persistent CSMA—applied to slotted channels. When the channel is found to be idle, a packet is transmitted in the first available time slot.

- CSMA/CD—in CSMA with collision detection (CD), a user monitors the transmission for collision. If two or more terminals start a transmission at the same time, the detected transmission is immediately aborted.

- Data sense multiple access (DSMA)—a special type of CSMA. The transmitter demodulates the forward control channel before broadcasting data in the reverse channel. Each user attempts to detect a busy-idle message. When a busy-idle message indicates no user is transmitting in the reverse channel, a user is free to send a packet.

2.6.12 Packet Radio

In the packet radio (PR) access technique, many subscribers are allowed to attempt to access a single channel in an uncoordinated manner. Transmission from each user is done in bursts of data. Simultaneous transmissions of multiple transmitters are detected at the receiver of the base station, which coordinates the operations by issuing acknowledged (ACK) and not acknowledged (NACK) signals to the requesting transmitters. The ACK signal indicates an acknowledgment of the received burst from a particular user by the same station, and NACK indicates that the previous burst was not received correctly by the base station. By using ACK and NACK signals, a PR system employs effective feedback, even though traffic delay due to collisions may be high.

Packet radio multiple access is fairly easy to implement, but has a low spectral efficiency and may result in serious time delays. The subscribers use contention techniques to transmit on a common channel. There are many protocols used in traffic coordination in PR systems. For instance, areal locations of hazardous atmospheres (ALOHA) family of protocols, developed for early satellite systems, are a good example of managing contention and collision. ALOHA allows each subscriber to transmit whenever they have data to send. The transmitting subscribers listen to the acknowledgment feedback to determine if transmission has been successful or not. If a collision occurs, the subscriber waits a random amount of time and then retransmits the packet. The advantage of the PR technique is the ability to serve a large number of subscribers with virtually no overhead.

2.7 Conclusion

This chapter discussed the fundamental principles of modern communication systems. The basics of electromagnetic wave propagation were explained. As the electromagnetic waves propagate, they encounter undesired effects such as losses, fading, reflection, refraction, and attenuation. The components necessary for RF communication systems were introduced. Fundamentals of digital communication technology were discussed in detail. Modern communication methods, modulation and multiplexing techniques, frequency spreading, and multiple access methods were detailed and some examples were given.

3

Data Transfer, Networks, Protocols, and Standards

Networking of hardware and software resources is essential to bring multiple devices together. Networking introduces efficiency by enabling the exchange of information, creating collaborative operations, and sharing the functions of equipment and devices. Networks are a collection of interoperational devices linked together by a communication medium and supported by suitable software. The software may be responsible for the functionality of part of the system or the entire system. A system is a group of interrelated parts with the focus on establishing an interrelationship between them to ensure efficiency, to facilitate integration of the application, and to share the resources.

Connecting devices together to form networks is a concept that has been used for many decades in a wide range of applications. In earlier networks, almost all the communicating devices were connected by wires, thus they were largely fixed in space. The devices in modern networks, as discussed in this book, can be interconnected with the use of wireless communication technology. Applications of wireless technology create mobility in space while still maintaining the network. Therefore modern networks can be viewed as wired networks in which the communication devices are connected by wires and are largely fixed in space; wireless networks in which devices communicate wirelessly and can move in space; and hybrid networks in which both wired and wireless techniques are used. At the moment, mobile networks based on wireless techniques provide primarily voice-based services, but these are increasingly handling data and other forms of information. Wireless networks can perform functions similar to those of fixed network, plus they offer many advantages such as reduced cost for initial setup and maintenance.

A great deal of commonality exists between wired and wireless networks. Wireless networks are built on top of the existing network technology, thus making use of the vast accumulated knowledge gained over many years.

In this chapter, the basic elements of networks and transmission of information are explained. The types of network topologies for wired and wireless media are discussed. Protocols and standards relevant to wireless networks are dealt with in detail, as they are essential for ensuring the interoperation of different equipment and devices. The concepts of security in wireless networks are highlighted and security methods are discussed. Because wireless technology continues to expand, the efficient use of the frequency spectrum and new emerging technologies are discussed.

3.1   Data Transfer

The term data refers to alphabetical, numerical, or special purpose characters that are appropriately grouped in binary form to constitute words, messages, or information. Data communication is primarily concerned with the transfer of data from a device in one location to a device in another location. Two or more devices communicating with each other form a system and the devices are said to be networked. Networks can be wired or wireless, or a combination of the two.

The transfer of data from one device to another is measured as the baud rate or bit rate. The baud rate indicates the number of symbols transmitted in a unit of time, usually per second. The bit rate indicates the number of bits transmitted per second. The baud rate and bit rate are the same only when one bit is allocated per symbol. But symbols are usually expressed as a series of bits forming words, streams, and codes. For example, Murray codes, which are used for numbers and alphanumeric characters, contain five bits per symbol.

Digital information is often transmitted in data frames. A data frame is a collection of characters conveying a complete message that can be understood by the transmitting and receiving devices. A typical data frame is shown in Figure 3.1. When data frames are used, the information rate is not the same as the bit rate or baud rate because it contains overhead data addresses, error checks, and starts and stops information. The type of information in the frames is governed by the protocols and standards used in that particular application. The protocols are configured within some reference model (e.g., Institute of Electrical and Electronics Engineers [IEEE] 802 and Open Systems Interconnection [OSI] reference models). Understandably, when protocols are used, the information rate may be much less than the quoted transmission rate.

Start bits	Destination address	Source address	Information	Frame check

FIGURE 3.1
Digital data transmission in frames.

The theory, the protocols, and the implementation of digital communication systems and associated networks that rely on physical connections such as wires or optical cables are well established and have been in use for many years. However, when compared to wired techniques, wireless data transmission and networking of instruments and sensors is relatively new and can offer many additional features. It is sufficient to say at this stage that the wireless components of most networks behave like their wired counterparts. Thus the operational principles of wired and wireless networks have many common points, but wireless networks are developing as a separate entity in technological developments and applications.

In both wired and wireless communication systems, data can be transmitted either in parallel or in serial forms; with synchronous or asynchronous information flow; or with simplex, half-duplex, or full-duplex data transmission modes. Therefore the discussions presented on these concepts are applicable to wired as well as wireless data communication.

3.1.1 Serial and Parallel Data Transmission

Data can be transmitted from one device to another in serial or parallel forms. In serial data transmissions, each bit of a code is sent sequentially, as shown in Figure 3.2. Consequently serial transmission can be achieved only by one pair of conductors connecting a receiver and a transmitter. In parallel data transmission, all bits or a number of bits of a code are transmitted simultaneously. Therefore the number of wires required equals the number of bits sent plus the return wire. For example, for an eight-bit code at least eight parallel wires must connect between the transmitter and the receiver, as illustrated in Figure 3.3.

3.1.2 Synchronous and Asynchronous Transmission

Serial data can be transmitted in two forms via asynchronous or synchronous transmission. Asynchronous transmission sends messages in blocks. This form of transmission may contain significant idle periods between blocks and is often used where high-speed data transmission is not required. Asynchronous data transmission uses data characters that contain information on the synchronization process, the nature and length of the data, and the locations of the first and last bit of the data block, so that the receiver knows

FIGURE 3.2
Serial Data Transmission.

Transmit, Tx   Receiver, Rx

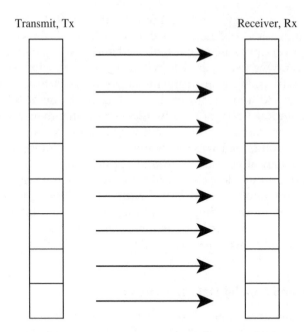

FIGURE 3.3
Parallel data transmission.

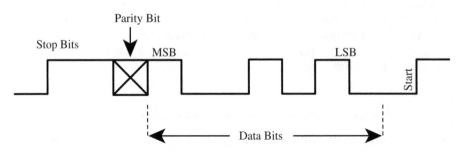

FIGURE 3.4
Asynchronous transmission of serial data.

the characteristics of the information coming from the transmitter. Since the receiver knows the start and stop bits of the block, the block can be sent at any time and at any rate. Each block between the transmitter and receiver is synchronized in its own right by the use of start and stop elements. The length of the data stream and the time gap between blocks are not usually fixed, but are decided on a per synchronization basis. Naturally, asynchronous transmission is slower than synchronous transmission because of the added synchronization needs.

Figure 3.4 illustrates a typical binary character transmitted in an asynchronous transmission form. When the character is transmitted, it is preceded by a start bit (binary 0) and followed by an optional parity bit and one or more stop bits. The stop bit is usually a mark or a binary 1.

In asynchronous transmission, the receiver detects the start bit by noting the transition from a mark to a space, and then decodes the next seven bits as a character. If more characters are to be transmitted, this process is repeated. The receiver and transmitter provide their own internal clocks, both being at about the same rate, but these clocks are not necessarily synchronized. Also, asynchronous transmission allows variable intervals between the transmitted characters.

Synchronous transmission is a message-based transmission technique; it does not employ start and stop bits as in asynchronous transmission. It requires a common clock pulse at the transmitting and receiving ends to achieve synchronization. The use of a common clock also helps character pulse identification. The receiver is able to recognize a unique code in the receive bits of the incoming data stream. This code allows the receiver to lock in the incoming bit stream. The receiver must be clocked at exactly the same clock rate as the transmitter. The synchronization of the clock is known as bit synchronization. Synchronous operation can be characterized as follows:

- There are no start and stop bits to synchronize each character.
- Every bit in the transmitter and receiver must be synchronized to a common clock.
- Data are sent in blocks that consist of many elements without separation in between.
- The entire block is framed by codes that indicate the beginning and end.
- The receiver must know the codes, the length of the block, and other relevant and control information.
- It is not sensitive to possible distortion of transmitted signals, since timing is done in a synchronized manner.

With synchronous transmission, synchronization is dealt with on a message basis rather than on a character basis. Once synchronized, it does not allow for a break or an interval between characters. This may limit the effective communication in devices that do not have continuous information flow or devices that do not have buffers to hold messages in case continuous transmission cannot be maintained.

Synchronous and asynchronous transmissions are handled by dedicated devices such as universal synchronous-asynchronous receiver/transmitters (USARTs) and universal asynchronous receiver/transmitters (UARTs). USARTs and UARTs are an important part of serial data transmission. A USART is a device that converts parallel bits into a continuous serial data stream, or vice versa. A USART can operate in synchronous or asynchronous form. A UART is a device that handles asynchronous serial communication. A typical UART is a 40-pin programmable device that transmits and receives asynchronous data in either half-duplex or full-duplex mode. A UART accepts parallel data and converts it into asynchronous mode to make it ready for a serial transmission.

3.1.3 Simplex, Half-Duplex, and Full-Duplex Data Transmission

The transmission of data between two devices can be characterized as simplex, half-duplex, and full-duplex, as shown in Figure 3.5. Simplex operation indicates that transmission can take place in only one direction from one device to the next. In this mode, one of the devices can transmit but cannot receive, or it receives but does not transmit. Half-duplex operation indicates transmission in either direction, but it can take place in only one direction at a time. Full-duplex operation indicates transmission takes place in both directions simultaneously.

In networks where many devices are involved in communication, transmission uses multiple channels. A channel is defined as a single path on a line through which signals flow. Lines are defined as the components and parts that extend between the terminals of the communicating devices.

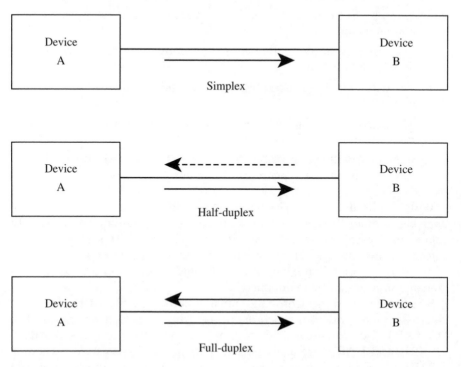

FIGURE 3.5
Three modes of channel operation.

An example of simplex operation is a paging system. In paging, messages are received but not necessarily acknowledged. An example of half-duplex operation is a walkie-talkie. In walkie-talkies, the operator pushes a button to talk and releases the button to listen. Both operators cannot communicate at the same time. A full-duplex system provides simultaneous, but separate channels by techniques such as frequency division duplexing (FDD) or time division duplexing (TDD). FDD uses different frequency channels and TDD uses adjacent time slots on a single channel.

3.1.4 Wireless Data Transmission

Devices can communicate by wired or wireless connections forming local area networks (LANs). The wireless components of most LANs behave like their wired counterparts, but they use space as the transmission media. The operational principles of wired and wireless networks are much the same: it is necessary to attach a network interface for the transmitting and receiving devices. In the case of wireless networks, the interface takes place mainly by radio frequency (RF) transceivers rather than cables. In many cases, wired and wireless systems are used in a mix-and-match form. When interfacing wireless systems to wired networks, devices called access points are used to connect both sides. This allows the shuttling of data traffic back and forth between the wired and wireless components.

In wireless communication systems, the frequency used for transmission affects the amount of data and the speed at which the data can be transmitted. The strength or power level of the transmission signal determines the distance over which the data can be sent and received without errors and corruption. In general, the principle that governs wireless transmissions dictates that a lower channel frequency can carry less data, more slowly, but over long distances. Although higher frequencies can carry more data at faster rates, the distance of effective transmission becomes shorter.

Modern wireless communication systems largely use the middle part of the electromagnetic spectrum. The middle part of the electromagnetic spectrum is divided into several frequency ranges, or bands, for communication purposes. These frequency bands are radio frequencies (10 kHz to 1 GHz), microwave frequencies (1 GHz to 500 GHz), and visible and infrared frequencies (500 GHz to 1 THz).

3.1.5 Radio Frequency Data Transmission

There is an inverse relationship between the frequency and the distance that an electromagnetic wave can carry data. There is also a direct relationship between the frequency and data transfer rate and the bandwidth. In data transfer, wireless networks make use of three primary frequency bands: RF (narrowband, spread spectrum, broadband), infrared and laser, and microwave.

Radio frequency communication systems are designed to operate as narrowband, spread spectrum, or broadband systems both in long-distance and short-distance operations. However, the use of this frequency band and the powers that can be transmitted at these frequencies are strictly regulated. In the United States, government agencies such as the Federal Communications Commission (FCC) regulate nearly all radio frequencies. Any commercial or government organization that wishes to use a particular frequency band must apply for permission. Permission may be granted to use that frequency for broadcasting in a specific location, usually with maximum transmission power limits. However, the FCC set aside certain frequencies for unregulated use within specified power levels for transmission. Typical unregulated frequency ranges are 902–928 MHz, 2.4 GHz, and 5.72–5.85 GHz, with some maximum broadcasting distances, typically about 70 m.

Narrowband radio or single-frequency radio networks use low-power and two-way radio communication systems in a half-duplex format. Radios in taxi cabs and base stations are good examples of such systems. In these systems, both the receiver and the transmitter must be tuned to a specific frequency to handle incoming and outgoing calls. Some single-frequency systems are made to operate at higher power ratings. Systems of this type can usually transmit over long distances and use repeaters and signal bouncing techniques to increase their coverage distance.

Spread spectrum radio systems address several weaknesses of single-frequency communication systems, both in high- and low-power operations. Spread spectrum uses multiple frequencies simultaneously, thereby improving the reliability and reducing the susceptibility to interference. Multiple frequencies make eavesdropping in data transfer much more difficult, if not impossible.

Two main types of spread spectrum communication are frequency hopping and direct sequence modulation. Frequency hopping is based on switching the data among multiple frequencies at regular intervals. The transmitter and receiver must be carefully synchronized to maintain communication. In many systems, hardware handles the timing of hops and chooses the next frequency for transmission.

Direct sequence modulation breaks data into fixed-size segments, called chips, and transmits the data on several different frequencies at the same time. The receiving equipment can identify the frequencies that are carrying data. Once the data is received from the identified frequencies, the receiver reassembles the arriving chips into properly arranged sequences of information as sent by the transmitter. For security purposes, some systems transmit dummy data on one or more channels along with the real data on another channel to make life even more difficult for eavesdroppers.

3.1.6   Infrared Data Transmission

Infrared wireless transmitters use light beams at infrared frequencies to send communication signals from the transmitter to the receiver. Infrared transmitters generate strong signals to prevent interference from other light sources. The communication systems work well mainly because of their high bandwidth. These systems can deliver data at speeds of 10 Mbps to 100 Mbps. There are four primary types of infrared systems:

- Line-of-sight systems require a clear line of sight, or an unobstructed view, between the transmitter and receiver.

- Reflective infrared system signals are generated by an intermittent device called the central hub. The central hub then forwards the messages to the intended recipients.

- Scatter infrared systems operate by bouncing the transmitted signals off of walls or other solid objects. The bounced signal is then picked up by the receiver. This approach limits the distance of transmission to typically 30 m or less, depending on the strength of the transmitted signal, the sensitivity of the receiver, and the presence of interference from other sources. Bounce technologies introduce delays in signal transmission, therefore scatter infrared systems operate on smaller bandwidths than line-of-sight systems.

- Broadband optical telepoint systems provide broadband services. Broadband systems operate with high-speed and wide bandwidths. In some applications, they can match the capabilities of most modern wired communication systems.

Laser-based transmission also requires a clear line of sight between the sender and receiver. In many applications, laser technology-based communication systems are subject to use limitations because excessive radiation can affect human vision and health.

3.1.7 Microwave Data Transmission

Microwave data transmission is an established technology that is used extensively worldwide. However, the infrastructure for microwave systems is expensive. Microwave data transmission may not be practical for small systems with short-distance data transfer, but the technology is available. It is used in aerospace communication, television broadcasting, and in military and some civilian applications for long-distance and high-rate data transmission.

3.2 Security in Data Flow

Security between the transmitter and receiver is very important in all types of communication systems. Data needs to be transferred without corruption (jamming) that can be caused intentionally by other parties and without being intercepted and listened to by third parties. Information delivered between the transmitter and receiver should be reliable, without any losses, erasures, additive noise, or fading, and it should not be intercepted by unauthorized parties.

There are many effective methods for the reliable delivery of information, such as channel coding, spread spectrum, multiplexing, and encryption. Channel coding is explained in this section. Spread spectrum and multiplexing techniques were explained in Section 2.5 through Section 2.7. Encryption is explained in Section 3.2.2. Reliable information flow also involves problems associated with electromagnetic wave propagation, such as multipaths and fading, as explained in Section 2.2. Since security in data flow involves a wide range of techniques, it is attracting a diverse range of research and development activities.

Security against the possibility of data interception by unauthorized parties is a great concern for both wired and wireless system users. The last thing a company wants is intercepted or compromised data. Theoretically anyone with an appropriate receiver positioned in the right place can eavesdrop. This is particularly true on narrowband, single-frequency systems. The use of encryption methods and the application of spread spectrum techniques makes eavesdropping and interception almost impossible.

3.2.1   Channel Coding

For successful information flow between two devices, the receiver must be able to recover the original signal from a received signal that might have gone through a number of changes during transmission. As discussed in Chapter 2, electromagnetic waves carrying messages are subject to many negative effects, such as fading, interference, channel perturbation, breaks in transmission links, and additive noise. Channel coding, also known as error control coding, is a method of protecting message signals from signal impairment by adding redundancies in the message signal, as illustrated in Figure 3.6. Channel coding can substantially reduce the probability of errors by guarding the data against the most probable errors or known errors that can occur during transmission.

In channel coding, the use of redundancy helps to distinguish the intended message even though there might have been significant corruption during transmission. The introduction of controlled redundancy creates subsets that

FIGURE 3.6
Concept of channel coding.

contain portions of the original message, thus, in a sense, hiding the message. The subset that contains the redundant portion is called the code and the valid messages are called code words or code vectors. A good code contains protected code words, so the likelihood of errors due to corruption during transmission is minimal.

The use of channel coding reduces the error probability as the redundancy in the code increases. However, adding redundancy increases the number of transmission bits from n bits to $n + k$ bits. This causes inefficiency in the system by reducing the transmission rate of the useful information. In applications, the codes are denoted by n and k, the code rate is given by n/k, and the increase in the data rate is $k/(n + k)$. Nevertheless, although errors are reduced, the use of channel coding does not guarantee the elimination of all errors.

There are two basic types of channel coding: block codes and convolution codes. Block coding partitions the source data into blocks of n bits. The encoder adds redundancy and converts the blocks to $n + k$ bits. The encoder also adds information on how redundancy is added, allowing the decoder, on the receiving side, to recover useful information from the codes. There are many types of error correction codes, the most well-known ones being Hamming codes, cyclic redundancy check (CRC) codes, Bose-Chaudhuri-Hochquenghem (BCH) codes, Reed-Solomon codes, and Goley codes.

Convolution coding is based on continuous operation of the encoder accepting useful data in blocks and using shift registers that generate sequences of higher rate data. Although this method is useful and convenient for error detection, error correction is much more complex than in block coding. Convolution methods employ techniques such as probabilistic decoding and approximation likelihood for error corrections.

Block and convolution coding techniques can be applied simultaneously, particularly in situations where channel errors occur in bursts. Both block and convolution coding use a technique known as interleaving, which spreads each message in a time interval and minimizes the effect of noise bursts. Interleaving sorts the data stream into a series of rows and applies coding in columns to find and eliminate errors. In this method, burst errors effectively become random errors so they can be handled as normal errors by error-correcting codes.

Once an error is detected, two main error correction methods can be applied: automatic repeat request (ARQ) and forward error correction (FEC). In ARQ, the receiver requests the transmitter to resend the part of the message that contains errors. ARQ is a powerful and effective technique, but it requires an additional feedback channel and adds delays in the data flow. In FEC, the receiver corrects the error without referring to the transmitter. It uses the additional information transmitted along with the data and employs one or more of the methods of channel coding.

In some communication systems, both random and burst errors can be severe and can occur simultaneously. In such cases, concatenation is used.

Concatenation uses two types of codes, one for correcting random errors and the other for correcting burst errors.

3.2.2 Encryption

Encryption is used for protecting the transmitted information from interception or corruption by unauthorized parties. Encryption converts the original text message into an encoded form, known as cipher text. When the data are encoded (encryption) at the transmitter, the resulting cipher text appears as a random stream of symbols that does not make sense. At the receiver, the encrypted data goes through a decryption process that recovers the original information. Encryption and decryption are both controlled by secret information, called the key, known only by the transmitter and receiver. The basic structure of the encryption and decryption process is illustrated in Figure 3.7.

Data are encrypted using ciphers, which are mathematical and physical processes for encrypting data. Ciphers can be altered or modified by changing the key that generates them. In conventional cipher systems, an identical key is used on the transmitter and receiver. This makes the encryption and decryption process symmetric. There are two major types of ciphers: the block cipher and the stream cipher.

A block cipher encodes a number of bits in predetermined blocks. A typical block size is 64 bits; that is, the original data of 64 bits are encrypted to form a cipher text block of 64 bits. A stream cipher encodes each bit individually in such a way that each bit of original text is converted to the cipher text. An ideal cipher text should be completely random and unpredictable. But this is not practical since it is difficult to synchronize the transmitter and receiver keys at all times during the transmission. Instead, most ciphers use pseudorandom key streams generated by the transmitter. The receiver is based on a shared key, which is the same key as the transmitter. The use of key streams makes the transmitted information appear as a completely random signal, thus there is virtually no observable relationship between the cipher text and original text.

Pseudorandom key streams can be generated in a number of ways. A convenient method of generation is the use of linear shift registers that consist of memory elements. The memory elements are arranged as shift registers, and the memory elements of the registers shift one position to the

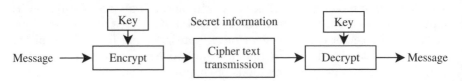

FIGURE 3.7
A typical encryption process.

right in each clock cycle. The output of the shift register is exclusive ORed with the original data. Since memory elements are only known to the transmitter and receiver, and the data are exclusive ORed with memory elements, an unauthorized receiver sees the stream of information as completely random and unrecoverable.

For an ideal encryption system, cryptanalysis, that is, recovery of the original text, should be impossible without knowing the correct key. Such a system is said to be unconditionally secure. However, the only known unconditionally secure encryption system is the one-time pad, in which the key stream is completely random and unpredictable, and is used only once between the transmitter and receiver. This type of unconditionally secure system is not feasible in most applications; instead, most systems aim to be computationally secure.

In computationally secure systems, the cost of breaking the encryption may be very high within the useful lifetime of the information. An attempt to break the encryption requires testing of every possible encryption key on the received blocks. This cost can be exemplified as follows: if a 56-bit key is used for encryption, all possible keys that can be used in the process will be 2^{56} keys. If, say, 10^{10} keys are tested every second, it would take about a month to test all the possible keys using a normal PC. One month is required to break the code of only one block, which makes it almost impossible to break the codes for a long stream of data. In general, the use of 128-bit or more keys provides long-term security within the operational speeds of today's computers.

An effective encryption system generates a cipher text so it is difficult to recover the original text if the encryption keys are not known. Ideally the cipher text should not have any observable structure. Thus the longer the key, the better the security. In many cases, minor changes in either the original text or the key can lead to larger changes in the cipher text. This is known as the avalanche effect.

There are many encryption algorithms and standards for wireless communication systems. The Data Encryption Standard (DES) is an example of a symmetric encryption algorithm. The DES was originally created by IBM. It is commonly used in Internet and banking transactions, and by cable television. Recently the DES was mandated by the U.S. government for use in securing data applications not involving national security.

There are many other popular encryption techniques. Asymmetric encryption algorithms are commonly employed. Public key and private key algorithms are two variations of asymmetric encryption techniques. The security in the public key algorithm is based on the differences in the complexities of some types of inverse operations. The Rivest-Shamir-Adleman (RSA) algorithm is probably the most popular public key encryption system. This algorithm uses two or more prime numbers and complex arithmetic operations for encryption and decryption. In practice, public key algorithms are often used at the transmitter and private key algorithms are used for decryption at the receiver.

3.3   Network Essentials and Topologies

Networks are arrangements of hardware and software components that communicate with each other in a coordinated manner. For effective communication, the components must be mutually compatible devices. Sharing resources and exchanging information among many users and devices on a network is called networking. The most elementary network consists of two devices that are connected together to transmit information from one device to the other. Even though the network concept appears simple, a great deal of coordination and many complex technologies are required to permit communication between devices. In addition, there are many possible choices for physical connections between the network elements and the associated software. This section discusses some of these complexities, thus giving the reader a better appreciation of network operations.

Networks are used for various reasons, including

- Data sharing, which permits a group of users to exchange information among each other on a periodic or regular basis.
- Routing data from one network element to another.
- Software resource sharing, which permits messages, documents, and other files to be shared by many users.
- Peripheral device sharing, which allows a group of users to share common hardware resources.
- Coordination of activities of equipment and devices in plant and production environments.
- Monitoring the status of equipment and devices.
- Monitoring and taking corrective actions on safety issues.
- Production planning, configuring, and reconfiguring.
- Research and development, pilot operations, unified system operation, etc.

There are various types of networks depending on the number of network elements and their spatial distribution. A local area network (LAN) is a system for interconnecting data communication components in a relatively confined space. LANs are most commonly contained within one or several buildings, as in industrial production facilities, universities, government departments, and other organizations. A networked collection of LANs is called an internetwork, as in the case of interdepartmental networks in universities and organizations. LANs can grow into wide area networks (WANs) that cover greater geographic distances, linking two or more separate LANS. In large, complex environments, the number of users and devices on a WAN can grow into thousands or more. For example, the Internet is a WAN internetwork that includes millions of machines and users worldwide. There are many other terms used to describe networks, such as metropolitan area networks (MANs), personal area networks (PANs), and so on, but these are basically LANs or WANs of various sizes.

The major purpose of networks is to share resources by connecting network elements, also called nodes. To be able to connect nodes, four elements are necessary: the transmission medium, the network topology, the protocols, and the network operating system.

Transmission medium can be defined as the physical path between the nodes of the network that connects the nodes to each other. The physical path may be cables, fibers, RF devices, microwave devices, etc.

Topology refers to the physical layout of the devices. Topology is linked to the communication methods used between the devices and the way that resources are shared. The network topology can have a significant effect on the performance and efficiency of the network, as well as its future growth potential.

Protocols are the set of rules that are agreed to that enable communication between devices. In the simplest case, for two devices to communicate with each other, they must share a common set of rules that clearly define how they will communicate.

The operating system is the software running in the background that manages the sharing of equipment and data between the network nodes. Operating systems are important because even though two devices might share a common medium and network protocols, they still may not be able to communicate with one another unless they run appropriate software to access the network and enable communication.

3.3.1 Network Software

Devices need network software to issue the requests and responses that allow them to communicate with each other. A communication process between two devices is illustrated in Figure 3.8. In this case, communication is taking place in simplex form: device A is sending information to device B.

In many networks, communicating devices invoke a layer of codes, which is called the network operating system (NOS). Network operating systems control access to network resources. Examples of common NOSs used in computers are Windows.NET, Windows XP, and Novell's NetWare.

Most network software packages come with modules for logging on and off the network. Network modules for logging on and logging off may include features such as password security, validation of user access to specific files and software, an automatic log on feature for some devices, help menus, error messages, and so on.

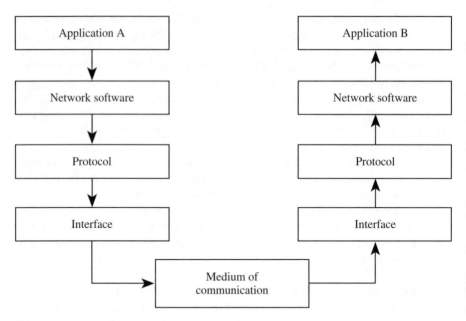

FIGURE 3.8
Process of communication in networked devices.

3.3.2   Network Topologies

Network topology refers to the physical layout of devices and supporting resources in a network and the communication methods between these devices. Topology has a significant effect on the performance of the network as well as on its future growth potential.

Topology primarily describes the patterns of connections of the network nodes. It determines the layout of the communication links between the terminal nodes, junctions, routers, repeaters, and servers. Topology algorithms are used for selecting links as well as link capacities for the associated devices on the basis of a number of factors, including transmission delays, costs associated with delays, the volume of total traffic, and future expandability of the hardware.

All devices, regardless of their topology, communicate in somewhat similar ways. They send data addressed to one or more recipients, transmit the data across the communication media, and accept and interpret the received data in a set way. In order to do all these, network elements obey some common protocols and standards shared between the involved devices.

Networks can be configured in different topologies or combinations of topologies supported by appropriate hardware and software. There are five basic types of topologies: bus, tree, ring, ad hoc, and star. Different network topologies are illustrated in Figure 3.9.

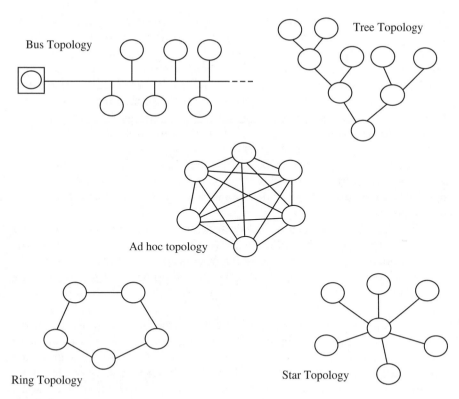

FIGURE 3.9
Different network topologies.

Bus topology consists of a single cabling arrangement to which all nodes are connected. The message is put on the bus by one of the nodes to be received by another node or nodes. Reception of the message is acknowledged only by the addressed node(s). Since all the nodes receiving the information are passive when there is a message on the bus, the system may be considered to be fail-safe. Each node can be individually installed, repaired, and disconnected without affecting the others. Faulty nodes can easily physically disconnected or isolated by software.

Tree topology is an expanded form of bus topology in which the cables branch in two directions, but offer only one transmission path between any two nodes. As in bus topology, any node can broadcast messages that can be picked up by any other node that is connected to the network.

Ring topology connects each node to two other nearby nodes by point-to-point links, forming a closed loop. Transmitted messages travel from one node to the next, going around the ring in one direction. When a device in the ring receives a message signal, it either acts on it by accepting the message or passes the signal to the next device in the ring. Each node recognizes the addresses of all other nodes and has an equal opportunity to send and receive information at any time.

Token passing is a method for sending data around the ring. In token passing, a node gains an exclusive right to use the channel by grabbing a token. By using the gained token, this node has the right to access the other nodes via the ring connections. When it finishes transmitting the information, it passes the token to another node that has data ready for transmission. When the intended destination receives the information, it returns a message to the sender acknowledging that the data has arrived safely. Each node may have a bypass mechanism that enables communication to continue if it is unable to pass information or goes down.

In some networks, a centralized controller acts as the main control unit and coordinates the other nodes to transmit messages. Modern ring topologies use smart hubs that isolate the faulty devices and ensure the flow of information.

In star topology, each node is connected via a point-to-point link to a central control node. All routing of network traffic takes place through the central node to the outlaying nodes. If a node wants to address another node, it has to go through the central node. Star topology has simple routing algorithms that mainly consist of lookup tables containing the addresses of all nodes.

In star topology, the central control node is the most complex of all the nodes. The complexity is governed by the efficiency, size, and capacity of the network. Star topology is desirable when the bulk of the communication takes place between the central node and outlying nodes. When the communication volume is high between the outlying nodes themselves, some delays may be encountered and efficiency may decrease.

One of the benefits of star topology is that it inherently centralizes the resources. Another benefit of star topology is the relative ease of troubleshooting. Nodes causing problems can easily be isolated by the central node without affecting the performance of other nodes. A drawback is that if the central node fails, all the devices attached to that node loose network access.

Ad hoc network topology is decentralized and does not rely on centralized and organized connectivity. Star, bus, tree, and ring topologies were primarily produced for wired systems, whereas ad hoc networks are most suitable for wireless networks in which a collection of autonomous devices communicate with each other. All network activity, including discovering potential nodes to communicate with, is executed by the nodes themselves. Once communication between the nodes is established, the nodes may organize themselves as one of the topologies (e.g., star or tree) explained above.

Ad hoc networks range from small, static networks that are constrained by power sources to large, mobile, highly dynamic networks. The design of network protocols for these networks is complex. Regardless of the application, ad hoc networks need efficient distributed algorithms to determine network organization, link scheduling, and routing. However, determining viable routing paths and delivering messages in a decentralized environment where network topology fluctuates is not a well-defined problem. In wireless systems, factors such as variable wireless link quality, propagation path loss, fading, multiuser interference, power expended, and topology changes become important issues. The network should be able to adaptively alter routing paths to alleviate any of these effects. Ad hoc networks are used in wireless instruments and this topology will be discussed in greater detail in Chapter 4 and Chapter 5.

3.3.3 Internetworking

Connecting LANs is called internetworking. Many LANs can be interconnected by using repeaters, bridges, routers, and gateways. Figure 3.10 illustrates LAN connection devices and their levels of operation in reference to the OSI model.

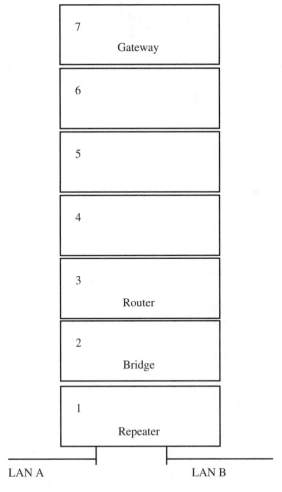

FIGURE 3.10
Interconnection devices for LANs and their levels of operation.

Repeaters are layer 1 devices. They are located between the transmitter and the receiver and their function is to strengthen the incoming signal and retransmit it, making it suitable for long-distance operations. Repeaters operate at the physical layer level, hence they do not understand or interpret the data frames or add any new functionality. Since repeaters do not have any intelligence, they can only be used to connect networks of the same type.

Bridges operate at layer 2 and can read the destination and source addresses embedded in the frame. Therefore bridges are able to redirect the frames to their intended destinations. Bridges cannot act like interpreters between different LANs, thus they can only interconnect networks of the same type. They are unable to convert frames if different network techniques are used in the internetwork.

Bridges have limited intelligence and can act as a filter by redirecting the frames onto a segment where the intended device is connected. This is a useful property for the efficient operation of LANs, since the volume of traffic and possible collisions are reduced. Bridges increase the throughput considerably, particularly in high-speed interconnected devices on similar LANs. This is particularly true in situations where protocol conversion is not required.

When the bridge receives a frame, it examines the source address and then it follows a set of rules in handling the frame. If the bridge knows the intended address, the frame is forwarded to that address. If the address of the incoming frame is not specified, the frame is ignored, resulting in the loss of information. If the address of the frame is not in the address table of the bridge, the frame is still forwarded. The bridge continuously updates the address table by adding new addresses of sending and receiving devices.

Bridges use layer 2 information to pass data to the correct destination. Some bridges, called router bridges or brouters, provide extra capabilities by being able to define WAN ports or performing conversion operations between different types of networks.

Routers are level 3 network devices. They are highly intelligent and have extensive knowledge of the networks they are involved in. Routers can determine the best route to the destination through the network, assign priorities to the information flow, and provide some limited security. Routers can link LANs that use different network protocols if they operate on a common protocol such as Transmission Control Protocol/Internet Protocol (TCP/IP).

Routers can also act as filters, as in the case of bridges, but they offer better network management capabilities. Because of these additional features, they are relatively slower. Selecting the best route between the source and destination is achieved by routing algorithms that consider complex network factors. An important network factor is the way that packets are handled. Depending on the volume of traffic, routers divide the long frames into packets, determine convenient routes for the packets, and manage the timing of those packets to send them to the destination to be assembled correctly as the original frame.

Gateways operate at layer 7 and are able to connect two or more totally different networks. They can act as translators between host machines within two networks. However, since gateways can route frames as well as act on them as translators and protocol converters, they are relatively slow devices.

3.3.4 Internet and Intranet

Internet stands for INTERconnected NETwork and represents the global "network of networks." The Internet is made up of a variety of computer networks that are connected through servers. Servers are computers that act as a center for connecting a particular LAN to the Internet. Servers are maintained by private and public Internet server providers (ISPs). A number of servers are connected to the Internet through gateways and routers. The complete network of the Internet is formed by the interconnection of thousands of routers. The transfer of information between computers that are connected to the Internet takes place through routers and gateways. Routers and gateways dynamically learn about network operations as they pass information from source to destination. Routers use packet switching techniques to pass information between computers.

The TCP/IP suite is a powerful tool used for communication on the Internet. TCP is a layer 4 protocol that provides services to applications by breaking down the messages into packets. IP provides management of packet switching.

Every computer connected to the Internet has its own discrete address—its Internet address or IP address. An IP address has four sets of eight-bit numbers separated by a period (e.g., 123.234.345.456). The first two sets of numbers identify the network, the next set identifies the subnetwork, and the last set identifies the computer in that subnetwork. IP addresses can be obtained from the ISPs, who have the rights to use block addresses obtained from an upstream registry.

IP addresses have a naming system that uses domain names rather than numbers. Specific documents on the Internet are located using a protocol called the Uniform Resource Locator (URL). For example, the address http://www.curtin.edu.au/mail locates electronic mail at Curtin University. In this address, http:// indicates the protocol or the access method, and www.curtin.edu.au indicates machine names and other domain names. The Internet name is followed by a slash, which indicates the directory and file names where the information that is requested resides. Apart from the Hypertext Transfer Protocol (http), there are other protocols such as the File Transfer Protocol (ftp) and Telnet.

The Internet is used for many applications, such as electronic mail (e-mail) transfer, file locations and display, bulletin boards for news groups (e.g., Usenet), file transfers (e.g., ftp), file organization (e.g., Gopher), Internet searches and explorations by browsers, and so on. The World Wide Web (www) is the most common application of the Internet. It is based on a client/server software model where the user is the client and www is the server. The World Wide Web is a system and collection of standards for storing, retrieving, formatting, and displaying information between computers on the Internet. World Wide Web documents are written in a language called Hypertext Markup Language (html) and transferred via http.

An intranet is a private Internet-like network set up by organizations that can be accessed only by members of the organization or authorized persons and groups. An intranet is established using the same architecture and standards (e.g., TCP/IP) as the Internet. Most intranets are connected to the Internet to provide authorized users with access to a wider range of resources. The connection of two or more intranets forms an extranet.

3.4   Protocols

As networks become more complex and wide reaching, the requirements for linking devices to networks and the interlinking networks themselves continue to grow. Thus the need for rules, regulations, and standards for successful connections increases proportionately. This leads to rules and standards that are accepted and practiced at national and international levels.

A protocol is a set of rules that are agreed to by relevant authorities to enable successful communication between devices. In the simplest case of two devices communicating with each other, they must share a common set of rules and procedures about how to communicate and exchange information. At this minimum level, such rules may include how to interpret signals, how to identify oneself and others on the network, how to initiate and end communication, how to manage the exchange of information across the network medium, and so on. Protocols must be comprehensive enough to regulate all essential requirements of a communication system. Some of the essentials requirements are

- Network topology: star, ring, bus, tree, or a combination.
- ISO reference model layers implemented: physical, data link, network, transport, session, presentation, and application.
- Data communication modes: simplex, half duplex, or full duplex.
- Signal type: digital, analog, or a combination.
- Data transmission mode: synchronous, asynchronous, etc.
- Data rate supported: from several bits per second to several gigabits per second, depending on the frequency of the operation and transmission medium.

- Transmission medium supported: wired, RF, optical, microwave, etc.
- Medium access control methods: carrier sense multiple access with collision detection (CSMA/CD), control token, etc.
- Data format: based on data transmission modes and individual protocol specifications.
- Error detection methods: parity, block sum check, CRC, etc.
- Error control methods: echo checking, ARQ, etc.
- Flow control methods: X-ON/X-OFF, window mechanisms, sequences, etc.

In communication systems, there are many different types of protocols addressing particular features of the communication process. Protocols are developed to enable communication in an entire network or part of a network, or in the communication devices within the network. For example, some protocols may be developed for connections only, some for transferring messages only, and so on. The majority of protocols are developed in a hierarchy of levels or layers in the OSI reference model. The establishment of a connection between two terminals may be realized by obeying the lowest few levels of the OSI model. The transfer of a file of information to solve a specific problem would follow higher levels on the hierarchy. The OSI reference model is explained in the next section.

3.4.1   The OSI model

Communication of devices in a network demands several carefully orchestrated activities and processes for the information to flow successfully between the sender and the receiver. The concept of networking activities and processes is as important as the configuration of the devices within the network. Several models had been proposed to create an intellectual framework within which to clarify network concepts and activities. Of all these models, none has been as successful as the OSI reference model proposed by the International Standards Organization (ISO). This model is commonly known as the ISO/OSI reference model, or simply the OSI model.

Because the OSI model is widely used and there are extensive hardware and software network elements complying with this model, it has become a key part of all types of communication and computer networking. An advantage of this model is that it makes many communication activities explicit by relating discrete activities and processes. The OSI model addresses all types of network concerns in an elaborate manner, starting from simple applications to completely open systems, and it has taken an unrivalled position in the world of networking.

TABLE 3.1

ISO/OSI reference model

Level	Function
Process control	Application and system activities control
Presentation control	Compacting, encryption, peripheral device coding and formatting
Session control	Support of session dialog
Data transport	End-to-end control, information exchange, reliability, error control
Network control level	Intranetwork operations, addressing and routing
Data link	Enables sequences to be exchanged across a single physical data link
Physical	Transmission at the physical medium

3.4.2   Structure of the OSI Model

The OSI model is configured in seven layers (Table 3.1). This layering approach helps to clarify the communication process for successful network operation. The layers are the essence of the OSI model. Essentially, networking can be broken into a series of related tasks, each of which can be conceptualized as a single entity in the communication process. This approach breaks down the complexity of the network, from the hardware supporting the network to the application software. Each individual task or activity is handled separately in each layer and issues concerning that layer can be solved independently. Once layers are developed independently, they can be interconnected to make a complete system with interrelated tasks and activities. This approach solves many problems by deconstructing the issues in each layer and breaking down overall system concerns into a series of smaller problems with possible individual solutions.

On top of the reference model resides the application layer, which provides a set of interfaces that permit applications such as Web browsers. At the bottom of the reference model, the physical layer resides. The physical layer is concerned with the networking medium, signals, physical connections, etc. All the activities necessary for successful network communication occur between the top and the bottom layers.

Data are typically transferred and interpreted between the transmitter and receiver, as illustrated in Figure 3.11. Layer construction follows a set process, allowing each layer on one device to behave as if it is communicating with its counterpart on the other device. This approach provides virtual communication between the peer layers. On the transmitter side, operations occur on the way down the stack starting from the application layer and moving toward the physical layer. On the receiving side, the role is reversed as the operations move up the stack.

At the transmitter, before data passes from one layer to the next on its way down the stack, it is broken down into protocol data units (PDUs), also called packets or payloads. The PDU is a unit of information passed

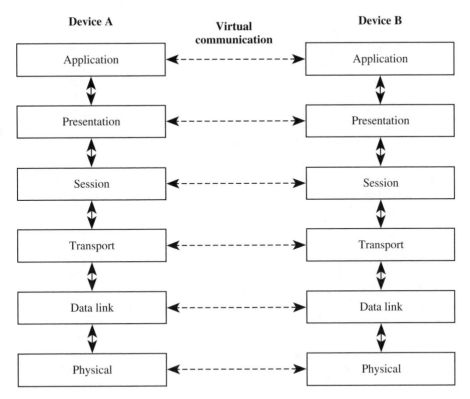

FIGURE 3.11
Relationship between OSI layers.

as a self-contained message to the peer as it moves from one layer to the next on its way down the stack. The protocol software adds its own payload information, formatting, and addressing to the PDU for delivery to its counterpart before it passes the packet to the next layer. An outgoing PDU generated by the sender in any given layer should substantially agree with the version of the PDU on the receiving side. This is important because the receiver should be able to recognize and interpret messages in the forwarded packet.

When the data arrive at the receiving end, the packet travels up the stack, starting from the physical layer through to the application layer. At each layer, the software reads its specific PDU data and interprets the message. Each layer strips its specific information and passes the PDU to the next higher layer. It also performs any additional processing if required.

For any device, be it sender or receiver, each individual layer of the OSI model has its own set of well-defined functions, and the functions of each layer communicate and interact with the layer above and below it. Each layer concerns itself only with the information exchanged between the peers at the sender and receiver. Each layer puts what may be considered "an electronic envelope" around the data it sends down the track for transmission; conversely, it removes the electronic envelope as it travels up the stack on the receiving side.

Rigidly specified boundaries, called interfaces, separate layers in the OSI model. Any request from one layer to another must pass through the interface. Each layer is built on top of the capabilities and activities of the layer below it and acts to support the layer above. Essentially each layer provides services to the adjacent higher layer and provides a shield for the above layer from the details of lower layers. All networked communication devices are equipped with compatible protocol stacks or protocol suits. Protocol suits are a collection of software elements and services that correspond to specific layers. Network access is only possible through the use of protocols and their associated drivers.

The functionality of the individual layers of the OSI reference model are discussed next.

Level 1 is the physical layer. All the details of creating physical network connections and the regulation of transmission techniques take place at this level. The physical layer divides the transmission data into frames and gets it ready for interfacing to the network medium. Protocols at this level involve parameters such as signal voltage swings and bit durations, the type of transmission (simplex, half duplex, or full duplex), how connections are established at each end, and so on. Thus the physical layer determines the way that streams of bits are translated into signals for transmission. A compatible receiving device accepts the signals and transforms them into a stream of bits to reconstruct the information. Encoding, timing, and interpretation of signals are decided at the physical level. A typical example of a physical layer protocol and hardware is the RS-232 standard used for serial communication.

Level 2 is the data link layer. In this layer, outgoing messages are assembled into data frames and acknowledgment from the receivers is awaited for each message transmitted. Data frames contain the identification (ID) of the sender and the receiver, as well as the controlling information. The destination ID provides a network address for the intended recipient and the sender ID provides the return address for return messages and acknowledgments. All outgoing frames include a destination address at the link layer, and if the higher levels require it, source addresses are included as well. Data integrity is checked at this level by the issued error detecting codes and error correcting codes. There are a number of techniques for data integrity checking, such as CRC, which is a special mathematical function based on the bit pattern in the outgoing frame. Information on error checking and correcting codes is sent as a part of the frame. Possible errors are determined on the receiving end by using special mathematical functions. If the mathematically determined values of error calculations agree with the sent values, the data are assumed to have been received in the original form. When the data are received at the destination, layer 2 protocols strip the information relevant to this layer and package the raw data to pass it on to the next layer. There are many examples of hardware and software operating at level 2 of the OSI model, such as IBM's Binary Synchronous Communications (BISYNC) protocol, X.25, and so on.

Level 3 is the network layer. The network layer handles addressing of messages for delivery and translates logical network addresses into their physical counterparts. Physical network addresses are known as media access control (MAC) addresses. MAC decides how to rout the transmission from the sender to the receiver. Level 3 considers important factors based on network conditions, quality of service information, the cost of alternative routes, and delivery priorities. It handles packet switching, data routing, and congestion control for the network. For the successful operation of this layer, long outgoing messages are divided into smaller packets for convenience in handling. Fragmentation and segmentation of data result in many packets, but provides easy manageability requiring shorter transmission times. When moving data from one type of network medium to another type, the network layer handles the segmented packets and helps reassemble the data that may have been altered due to dissimilar media. Incoming packets are reassembled into messages and downsized into their original forms to pass them on to higher layers.

Level 4 is the transport layer. The transport layer manages the conveyance of data from the sender to the receiver across the network. An important task of this layer is to ensure flow control by making sure that the recipient of the transmitted data is not overwhelmed with more data than it can handle. Therefore long data payloads are segmented into portions matching the maximum packet size that is acceptable to the network medium and the recipient. The transport layer of the receiving side resequences the divided data arriving in packets into its original form. The transport layer determines the necessary parallel paths and multiplexers for routing of the packets through available channels. The transport layer is the busiest layer for end-to-end communication in the network.

Level 5 is the session layer. The session layer establishes system-to-system connection across the network and allows two parties to carry out ongoing communication, called a session. The exchange of messages and the transmission of data takes place as long as this session continues. This layer has many functions, including setting up the session, monitoring the session identification process, providing security in data flow, providing continuity in exchanging data and messages, and terminating the session when the task is completed. The session layer also ensures synchronization of the tasks on both ends of the connection. It can place check marks in the data stream so that if communication fails at some point, only data after the most recent check mark is retransmitted. It controls logging on and off the system, verifies user identification from lookup tables, and provides billing and management issues.

Level 6 is the presentation layer. The presentation layer handles data formatting to make it suitable for networked communication. For outgoing messages, the presentation layer converts data into generic formats so that they can survive the rigors of transmission on the network. For incoming messages, it converts the data from its generic network representation into a format that is suitable for the requirements of the receiving device. The presentation layer performs many other tasks such as protocol conversion, handling library routines, performing data encryption and decryption, addressing character set issues, conducting compression and decompression, and providing graphic commands and code conversions.

Level 7 is the application layer. The application layer is the top layer of the OSI reference model. This is the level seen by individual users, and it provides a set of interfaces for applications to gain access to network services. In the application layer, network transparency is maintained by concealing the physical distribution of resources from the user. It provides service support applications for file transfer, message handling, database management, and so on. The application layer is able to partition complex problems among several machines in distributed process applications.

3.4.3 IEEE 802 Network Model

The IEEE 802 network model is based on the OSI reference model, but it is perceived as an enhancement of the OSI model. IEEE 802 lays out a family of specifications for different types of networks, consequently there are many protocols and standards within this model. The specifications of the protocols and standards encompass and meet the requirements of a diverse range of existing networks and they are conceived to be open-ended, thus allowing the development of new types of networks. The IEEE 802 standards are the most influential networking standards in use internationally.

IEEE 802 expands the OSI reference model at the physical and data link layers. At the data link layer, two additional sublayers are included: logical link control (LLC) and media access control (MAC), as shown in Figure 3.12.

The LLC sublayer controls data link communication and defines the use of logical interface points, called service access points (SAPs). LLC is also responsible for error recovery in some applications. There are several modes of LLC operation; some modes require LLC to detect and recover from errors that have taken place during transmission in the selected media.

The MAC sublayer provides shared access of multiple devices in the physical layer. MAC directly communicates with internal operations of a particular device, such as computers equipped with network interface cards (NICs), and is responsible for ensuring error-free data transmission between the device and the network.

The IEEE 802 model concentrates on standards that describe the physical elements of a network, including network adapters, cables, connectors, signaling technologies, MAC, and so on. Most of these reside on the lower two layers of the OSI model, in the physical and data link layers. IEEE 802 is also concerned with how to manage, attach, and detach devices in and out of a networked environment.

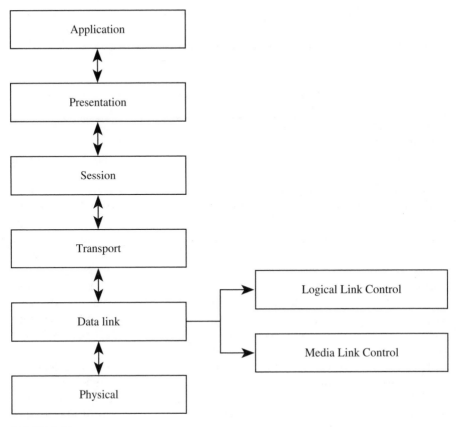

FIGURE 3.12
IEEE 802 with data link sublayers.

3.5 Standards

Standards are worthy of explaining in detail, as there may be a confusing number of standards for similar functionality devices, communication systems, and networks, as in the case of industrial fieldbuses as explained in Chapter 1. The term "standard" has many definitions. The *Oxford English Dictionary* defines standard as a noun, "1: a level of quality or attainment, 2: a required or agreed level of quality or attainment, 3: something used as a measure, norm, or model in comparative evaluations." Standard as an adjective, "1: used or accepted as normal or average, 2: (of a size, measure, etc.) regularly used or produced." From these definitions, the word "standard" implies comparative evaluations, which can be physical quantities and/or processes and procedures. The National Standards Policy Advisory Committee (http://www.nist.gov/) describes a standard as "a prescribed set of rules, conditions, or requirements concerning definitions of terms: classification of components; specification of materials, performance, or operations; delineation of procedures; or measurement of quantity and quality in describing materials, products, systems, services, or practices."

Since the word standard is a broad term and used in many contexts, there are several kinds of standards. In addition to the fundamental physical measurement standards, there are published standards (sometimes referred as paper standards) of practices and protocols that can be defined as documents describing the operations and processes to achieve unified results. Both physical and published standards play fundamental roles in shaping the efficiency of domestic and global economies. Published standards are important as they are documents that provide textual and illustrative information on what and how things should be done and are done.

Because of the national and international implications, there are many institutions and organizations that are responsible for investigating, developing, determining, and maintaining the relevant standards to support worldwide scientific and industrial activities. However, national and international standards authorities are subject to various internal and external forces, differences of opinion, and commercial interests during the development of standards and may not always end up with ideal results in the process of standardization. This, unfortunately, can result in a confusing number of standards and de facto standards with different versions and interpretations describing the same process. In addition, publications grow over time, and as new technological developments occur, the standards authorities and procedures change. This can result in a multitude of standards on a specific issue.

A standard is intrinsically a bureaucratic process that becomes obsolete in fast moving technological environments. Standards largely benefit users (as in the case of mobile communication) and manufacturing organizations. Vendors support standards to penetrate into a market and maintain their market position by offering interoperable or interchangeable components, devices, systems, and software.

It is important to realize that standards are developed by various national and international bodies, sometimes operating totally independently, hence there may be different versions for the same subject. Some examples of national and international standards bodies are the ISO, International Electrotechnical Commission (IEC), IEEE, and American National Standards Institute (ANSI). This list can grow to hundreds of organizations worldwide. All these organizations have multiple internal departments, committees, subcommittees, and working groups to support their activities.

The evolution and expansion of wireless technology is a dynamic and ongoing process, thus there are very few standards in this area as the technology is continuing to change and develop. Nevertheless, some key developments have occurred in the standards that are directly applicable to wireless instrumentation systems. A few examples of these relevant standards are

TABLE 3.2

IEEE 802 wireless network standards

Standard	Name	Function
802.1	Internetworking	Routing, bridging, and Internet work communication
802.2	Logical link control	Error control and flow control over data frames
802.3	Ethernet LAN	Forms of Ethernet media and interfaces
802.4	Token bus LAN	Forms of Token bus media and interfaces
802.5	Token ring LAN	Forms of Token ring media and interfaces
802.6	Metropolitan area networks	MAN technologies, addressing and services
802.7	Broadband Advisory Group	Broadband networking media, interfaces, equipment
802.8	Fiber-optic Advisory Group	Fiber-optic media, network types, technologies
802.9	Integrated networks	Integration of voice and data traffic in a network medium
802.10	Network security	Network access control, encryption, certification, other security
802.11	Wireless networks	Wireless networks, frequency usage
802.12	High-speed networking	Variety of 100 Mbps-plus technologies
802.13	Unused	
802.14	Defunct working group	Data transfer over cable TV
802.15	Wireless personal area networks	Emerging standards for wireless PANs
802.16	Wireless metropolitan area networks	Wireless MANs
802.17	Resilient packet ring	Very high speed ring-based LANs and MANs
802.18	Wireless Advisory Group	Radio-based wireless standards

IEEE 802, high performance radio local area network (HiperLAN), PAN, Bluetooth, cellular packet radio standards, and IEEE 1451.

3.5.1   IEEE 802 Standards

The IEEE 802 and ISO/OSI models were developed in collaboration of the organizations and are compatible with one another. IEEE 802 goes into much more detail on various types of networks, internetworking, high-speed networking, and network security. The network standards are numbered from 802.1 through 802.18. Each standard may have a series of standards carrying the same number but different extensions, such as 802.11b, 802.11g, etc. A complete list of 802 standards is given in Table 3.2.

The IEEE 802.11 Wireless LAN Working Group was founded in 1987 to begin standardization of spread spectrum wireless local area networks (WLANs) for use with industrial, scientific, and medical (ISM) bands. WLAN efforts of the IEEE did not gain momentum until the late 1990s, when the popularity of the Internet, combined with the wide-scale acceptance of portable devices and laptop computers, caused WLAN to become an important and rapidly growing segment of modern wireless communication. IEEE 802.11 was standardized in 1997, with the goal of providing interoperability standards for WLAN manufacturers using 11 Mbps direct sequence spread spectrum (DSSS) spreading and 2 Mbps user data rates. With an international standard now readily available for everyone, numerous manufacturers began to comply and the market began to grow rapidly. In 1999, the 802.11 high rate standard, called 802.11b, was approved, thereby providing new user data rate capabilities of 11 Mbps and 5.5 Mbps.

The IEEE 802.11 (simply 802.11) standards address issues concerning wireless networks. 802.11 will continue to be developed and grow because of new wireless technologies and applications. Many manufacturers of wireless networking devices and systems have developed inexpensive, reliable wireless LANs and associated devices for domestic and industrial use that comply with the 802.11 standard. There are several versions of the 802.11 standard, the current ones being 802.11b, which specifies a bandwidth of 11 Mbps at a frequency of 2.4 GHz; 802.11a, which specifies a bandwidth of 54 Mbps at a frequency of 5 GHz; 802.11g, which specifies a high-speed wireless standard operating at speeds of up to 54 Mbps with a carrier frequency of 2.4 GHz. These 802.11 wireless LAN standards are commonly known as Wi-Fi.

3.5.2   Wireless Ethernet Concepts

802.11 wireless networks are viewed as an extension of Ethernet that uses electromagnetic propagation as the medium of transmission instead of electrical and optical cables. However, most 802.11 networks incorporate some wired Ethernet segments operating collaboratively with the wireless components. The communication range of 802.11b-compatible devices is short, about 100 m, but the range of the network can extend from several meters to several hundred meters with the use of repeaters and other assisting devices, depending on the environmental factors and RF interference that exists in the area of operation.

802.11b uses a wireless access point (WAP) that serves as the center of a network configured in star topology. Workstations equipped with wireless devices such as network interface cards (NICs) can send packets to the WAP, which then redirects the packets to a destination workstation.

Wired Ethernets use CSMA/CD as the access method, but wireless networks have a special problem with this method. CSMA/CD requires that all the stations hear each other so they can identify the source that is sending data. If any two stations in the network try to send data at the same time, a collision can occur. In wired networks, the sending station will notice that a collision has occurred and it will attempt to send the data again. However, 802.11b wireless stations cannot send and receive data at the same time, so if a collision occurs, it may not be detected by the sending station. For this reason, 802.11b specifies a CSMA/CA access method in which an acknowledgement is required from the receiver for every packet that is sent and received. If there is no acknowledgment, the sending station knows the packet did not arrive safely.

A problem in wireless networks is the hidden node problem. Some stations may be out of range of each other and thus will not be aware that they are sending data at the same time. To counteract the hidden node problem, 802.11b specifies an optional feature that requires handshaking before transmission takes place. In this implementation, a station must send the WAP a request to send (RTS) packet requesting permission for transmission. If all is clear, the WAP sends a clear to send (CTS) message and the sending station can begin the communication process.

The 802.11b standard specifies a transmission rate of 11 Mbps, but adverse environmental conditions may prevent transmission at this speed. Therefore transmission speeds may be decreased incrementally starting at 11 Mbps to 5.5 Mbps to 2 Mbps and finally to 1 Mbps for reliable connections. In the 802.11b standard, there is no fixed segment length because reliable communication depends heavily on the environment and the segment length is determined by environmental conditions.

In general, an 802.11b network has a maximum range of about 100 m with no obstructions. This distance can be extended using large, high-quality antennas. However, the data rate may suffer as the distance increases or as more obstructions are present in the transmission path.

The 802.11g and 802.11a standards are extensions of 802.11b. Although 802.11g competes with 802.11a, it shares many common features. 802.11g specifies a bandwidth of 54 MHz, while 802.11b specifies 11 MHz. The 802.11a standard uses an unlicensed 5 GHz portion of the spectrum, but at this time 802.11a products are more expensive to produce. 802.11g is backward compatible with 802.11b.

The 802.11a frequency bands are 5.15–5.35 GHz and 5.725–5.825 GHz, allowing at least eight simultaneous channels. In addition, another band of 255 MHz will be available for 802.1a in the United States at 5.47–5.725 GHz. 802.11b and 802.11g use an 83.5 MHz band located between 2.4 GHz and 2.4835 GHz, allowing three channels to be used simultaneously.

Of these competing standards, 802.11b appears to be most prevalent, but it has been in use longer. Of the two higher speed standards, 802.1g is backward compatible with 802.11b and therefore provides convenient bandwidth upgrade possibilities. 802.11a presents problems for upgrades to 802.11b because of its higher frequency, but it is far more reliable and flexible. It is likely that standards that provide high-speed transmission will ultimately dominate the marketplace and will be accepted by consumers as well as manufacturers for widespread use.

3.5.3 IEEE 802.16 Wireless Metropolitan Area Networks

The IEEE 802.16 wireless metropolitan area network (WMAN) standards are equivalent to HiperMAN in Europe. It can be seen as a replacement for wired or fiber-optic-based MANs. 802.16 forms the backbone of many wireless Internet service providers (WISP) operating around the globe. The base station is the fundamental component of a WMAN. The base station serves as a hub and can easily be located on buildings or transmission towers. Base stations can transmit information over long distances, typically 10 km to 15 km. Base stations are also used as bridges between the wireless world and the Internet. IEEE consortiums such as the Wi-Fi alliance try to ensure that 802.16 and 802.11 products are compatible with each other. Some WMANs use the unlicensed 2.4 GHz frequency band and orthogonal frequency division multiplexing (OFDM) techniques for data transmission.

3.5.4   Code Division Multiple Access-Based Standards

Code division multiple access (CDMA) is a cellular technology that competes with other technologies such as the Global System for Mobile Communications (GSM), Digital Enhanced Cordless Telecommunications (DECT), and General Packet Radio Service (GPRS). CDMA, developed by Qualcomm and Ericsson, is a high-capacity and small cell radius cellular system. It employs spread spectrum technology and a special coding scheme. The first generation of CDMA is known as CDMA One (also known as IS-95), and the second generation is CDMA2000, which is the dominant technology at the moment. CDMA2000 has many variants such as 1X EV, 1X EV-DO, and MC 3X. New versions of CDMA are emerging, as the third generation, in the marketplace; some examples are Multicarrier CDMA2000 1xMC and HDR (high data rate) operating in the 1.25 MHz bandwidth, and 3xMC and Direct Spread WCDMA operating in the 5 MHz bandwidth.

CDMA2000 is a wideband radio interface that offers significant advances over CDMA One with its improved performance and capacity by using turbo codes. CDMA2000 employs advanced MAC for efficient and high-speed packet data services. Its physical layer features a dedicated control channel (DCCH) and a common control channel (CCCH). The specifications of CDMA2000 also incorporate advanced multimedia quality of service (QoS) capabilities to enable scheduling and prioritization among competing services.

3.5.5   Time Division Multiple Access-Based Standards

Time division multiple access (TDMA) standards use a single channel and divide it into a number of time slots. Each user is allowed to use only one time slot out of every few slots. Some network systems use dynamic time slot allocation to avoid wasting time slots if one side of the conversation is silent.

Time division multiple access was first implemented as the TIA-54 standard (also known as D-AMPS). TIA-54 provided three TDMA voice channels in the space of one 30 kHz analog channel. The next generation of TDMA, known as IS-136, extended the use of TDMA to the control channel. IS-136 has been adapted by ANSI and published as the TIA/EIA-136 series of standards. The UWC-136 RTT standard is based on an enhancement of the ANSI-136 TDMA standards that incorporate several others such as GSM, Enhanced Data GSM Environment (EDGE), and GPRS. It includes enhanced interfaces, capabilities, features, and services. A greater spectral efficiency is achieved with a change in the modulation scheme, new slot format, the addition of new interleaving and coding options, and other enhancements.

Time division multiple access, as defined in TIA IS-54, IS-136, and TIA/EIA-136, divides the 30 kHz cellular channel into 3–8 kbps time slots, which supports three users in strict alternation. This approach theoretically triples the capacity of cellular frequencies.

The current version of ANSI-136 supports 30 kHz channel spacing, six time slots, three calls per channel, circuit switch data rates of up to 28 kbps, and more. The UWC-136 compliant version of ANSI-136 supports 30 kHz, 200 kHz, and 1.6 MHz channel spacing, six time slots with either six or three calls per channel, and high-speed packed data rates up to 384 kbps or 2 Mbps.

3.5.6 GSM and GPRS Standards

The GSM is a digital cellular system that has found popular applications around the world. GSM networks operate in three different frequency ranges: GSM 900 operates at 900 MHz and is most common in Europe and rest of the world; GSM 1800 or DCS 1800 operates at 1800 MHz and is used in many countries (e.g., France, Germany, Switzerland, United Kingdom, and Russia); and PCS 1900 or DCS 1900 operates at 1900 MHz and is used in the United States and Canada. Apart from their operational frequencies, the only differences between these systems are power levels and some minor changes in signaling. GSM standards are used for providing interfaces between various entities in GSM networks. There have been many standards introduced recently to support GSM networks. Some of these standards are GSM 04.31 (v. 8.1.0) for radio resource protocols, GSM 08.71 (v. 7.2.0) for base station system interface layers, and many others.

GSM has many features that can be used in instrumentation systems and networks, such as the short message service of GSM, which allows sending and receiving of 126-character text messages. The data speed is 9600 bps and it uses effective encryption techniques to prevent tapping and eavesdropping.

GPRS is based on GSM. It is a packet-based wireless communication service that has data rates of 56 kbps to 114 kbps. Packet-based services are more efficient than circuit-switched services since communication channels are being shared between users rather than dedicated to only one user at a time. GPRS allows several users to share the same GSM time slots via a link layer send/receive scheduling protocol.

GPRS complements Bluetooth, which effectively replaces wired connections between devices with wireless radio connections. GPRS also supports IP and X.25, a packet-based protocol that is used mainly in Europe. Hence, with GPRS, the user can access two forms of data networks, X.25 for packet-based systems and IP.

In GPRS, a TDMA packet data channel (PDCH) carries both user data and information. GPRS has a series of standards, such as GPRS-136. The main goals of these standards are to provide network architecture, radio communication technologies, and protocols for access, such as for roaming between GPRS and other networks. The GPRS-136 data model overlays the circuit-switched network nodes with packed data network nodes for service provisioning, registration, mobility management, and accounting. GPRS is an evolutionary step toward EDGE and universal mobile telephone service (UMTS).

3.5.7 Other Wireless Network Standards

There are many other wireless network standards such as composite CDMA/TDMA, Digital Advanced Mobile Phone System (D-AMPS; also called IS-54), DECT, EDGE, GSM EDGE Radio Access Network (GERAN), iMode, Personal Communications Service (PCS), Personal Digital Cellular (PDC), UMTS, Worldwide Interoperability for Microwave Access (WiMAX), and so on. These standards will not be explained here. Please consult the references for more information.

3.5.8 IEEE 1451 Standards for Smart Sensor Interface

There are new standards that are emerging for hardware architecture, software, and communication of modern intelligent (smart) sensors. These standards are making a revolutionary contribution to wireless instruments, instrumentation, and networks.

In traditional instruments, sensor output, largely analog, is processed further for measurement and display purposes. Incorporating microelectronics and microprocessor technologies in both sensors and instruments has increased their functionality. In addition, new technology has emerged where sensor signals can be directly interfaced and processed without any dedicated circuits on digital platform. With integrated communication capabilities, networks that talk directly to the sensors have emerged. Sensor networking is becoming pervasive and is causing a major shift in the measurement arena.

The rapid development and emergence of smart sensors and the associated networking technologies is making smart transducers an economical and attractive solution in many measurement and control applications. However, the existence of many incompatible networks and protocols makes it very difficult to interface a wide range of sensors. In addition, a sensor customized to interface with a particular network will not necessarily work with other networks. It is clear that no particular control network is becoming the industry standard at this point in time. It appears that a variety of networks will coexist to serve specific industries. Many manufacturers are uncertain of which networks to support and they are holding back on full-scale sensor production. This condition impedes widespread adoption of smart sensors and networking technologies, despite the compelling desire to build and use them.

In view of the situation, the Instrumentation and Measurement Society of the IEEE has set up the Sensor Technology Technical Committee to organize a series of workshops to provide an open forum to exchange ideas on sensor interface issues. As a result, a series of IEEE 1451 standards were proposed and accepted.

The objective of the IEEE 1451 standard for a smart transducer interface is to define a set of common communication interfaces for connecting transducers to a microprocessor-based system, to an instrument, or to a field network in a network-independent environment. IEEE 1451 is a set of standards that defines interfaces for network-based data acquisition and control of smart sensors and transducers. The aim of the IEEE 1451 is to make it easy to create solutions using existing networking technologies, connections, and common software architectures. The standard allows application software, field network, and transducer decisions to be made independently, thus providing flexibility in choosing products and vendors that are most appropriate for a particular application.

The ultimate goal of the IEEE 1451 standards is to provide some means of achieving transducer-to-network interchangeability and transducer-to-network interoperability. To achieve this goal, these standards provide clear ways of creating measurement and control devices at the process connection level. Sensors complying with these standards are expected to have onboard information on serial numbers, calibration factors, accuracy specifications, and so on. During installation, location information can also be loaded.

The family of IEEE 1451 standards can be divided into two basic parts:

- Defining a set of hardware interfaces for connecting transducers to a microprocessor or instrumentation system.

- Defining a set of software interfaces for connecting transducers to different networks while utilizing existing network technologies.

Understandably the IEEE 1451 standards cover wide spectrum of applications. IEEE 1451 is divided into six sections—1451.1 to 1451.6.

IEEE 1451.1 is the first section and is known as the Network Capable Application Processor (NCAP) information model. This section is concerned with the software architecture that moves the intelligence to the device level. The 1451.1 standard uses object modeling to describe the behavior of the smart transducer. It supports transducers by way of a series of transducer blocks, which can be viewed as the input/output (I/O) driver abstraction of hardware. The application software that supports the operation of smart transducers can access the transducers through the application programming interface (API). The API is a set of standardized software function routines such as "IO_Read" to request a specific type of transducer electronic data sheet (TEDS) data, and "IO_Control" to set and reset sensor parameters. The software creates a flexible environment and natural modules that allow engineers to think at the level of operational real-world systems, not at the level of programming languages. This type of approach creates object-oriented technology that makes open systems possible. Object-oriented systems produce devices that are much easier to adapt to new application demands. In this way, flexibility can be achieved so that systems can be assembled, reassembled, or modified quickly and transducers can easily be reconfigured and connected to different networks.

IEEE 1451.2 is the second section and it is concerned with transducers, microprocessor communication protocols (MCP), and TEDS. The IEEE 1451.2 standard defines the transducer data and electronic interface of digital information directly from the sensor to the system, thus creating a modular type of architecture. This modular architecture allows embedding of modules in any field network in an automatic and transparent manner. This section proposes a 10-wire interface, called the transducer independent interface (TII). Two of the wires are power and ground, other lines are allocated to data in, data out, and a clock. A dedicated sensor detect line allows the NCAP to determine if a sensor is plugged in for plug-and-play operations. The interrupt line allows the Smart Transducer Interface Module (STIM) to request service. The trigger line initiates a sensor measurement and acknowledges completion of the requested action.

IEEE-1451.3 is the third section and mainly looks after the digital communication and TEDS formats for distributed multidrop systems.

IEEE 1451.4 is the fourth section and is concerned with mixed-mode communication protocols and TEDS formats. IEEE 1451.4 establishes a universal system for the data that digital networks need to identify, characterize, interface with, and use for signals from analog sensors. This section aims to simplify the installation, creation, and maintenance of sensor networks. TEDS formats specified in the standard are self-identifiers, which are usually placed on chips embedded in sensors and actuators. Each TEDS node supplies the data a network needs to identify a device and interpret what is in its memory. TEDS formats are written in the Template Description Language described in IEEE 1451.4. The standard defines templates for commonly used devices such as accelerometers, strain gauges, microphones, and thermocouples.

Any company can apply to IEEE to register their devices. IEEE allocates unique registration numbers (URNs) and manufacturer identifiers for use in TEDS formats. Each URN has an 8-bit family code to designate the type of device, a 48-bit serialization field to identify the individual device, and an 8-bit redundancy check code to verify communication integrity. Transducer manufacturers install a 64-bit URN in each unit they make. Serial numbers

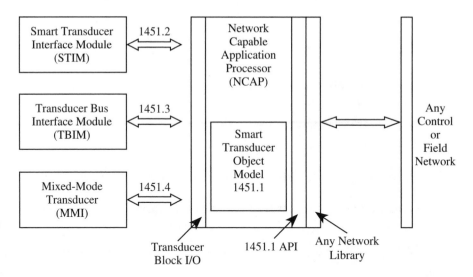

FIGURE 3.13
The relationship of the IEEE 1451 family of standards.

for URNs are bought in blocks of 4096 numbers from the IEEE Registration Authority.

IEEE 1451.5 is the fifth section and is concerned with wireless communication of sensors. This section specifies information that will enable 1451-compliant sensors and devices to communicate wirelessly. The IEEE is currently working on three different standards—802.11, Bluetooth, and Zigbee.

IEEE 1451.6 is concerned with the information required from devices to operate on consolidated auto network (CAN) buses.

Figure 3.13 illustrates the relationships between different sections of the IEEE 1451 standard. IEEE 1451.1 deals with software. The other sections define the physical interfaces for smart transducer connectivity to networks as well as TEDS and its data format.

There are positive signs that many companies are complying with the IEEE 1451 standards. For example, National Instruments is actively promoting and introducing plug-and-play sensors that comply with IEEE 1451.4. There are many smart sensors available that comply with these standards. More information on these sensors and the IEEE 1451 standards will be provided in Chapter 4.

3.6 Wireless Networks, PANs, LANs, and WLANs

Wireless data transmission and networking is particularly appealing to many users because it eliminates wires, reduces costs of installation and maintenance, and provides mobility. In recent years, wireless technology has attracted considerable commercial as well as research and development interest. Wireless networks and related technologies provide the following:

- Easy integration of compatible devices to existing wireless networks.
- Connection to the Internet.
- Mobility of users and devices.
- Interoperability and system integration.

Today, wireless technologies and networks are applied in many areas:

- Networking of instruments in plants and factories.
- Wireless instrumentation and control of industrial processes.
- Wireless networking in environments that are constantly changing, such as in wildlife and habitat monitoring.
- Simple messaging between fixed and mobile devices.
- Connection to the Internet of mobile devices or users.
- Advanced wireless Internet access.
- Delivery of network access into isolated areas.
- Improved customer service in busy public areas such restaurants.
- Network connectivity in facilities where wiring would be prohibitive and costly.
- Home networks where the installation of cables is inconvenient and expensive.
- Secure communication, such as corporate LANs.
- Environmental monitoring.
- Health industry and patient monitoring.
- Machine health and diagnostics.

Current wireless technologies and associated networks consist of a layered combination of different access technologies, including

- Wireless LAN (e.g., IEEE 802.11a, 802.11b, HiperLAN, etc.).
- Personal LANs for short-range and low-mobility applications (e.g., Bluetooth, Infrared Data Association [IrDA]).
- Cellular systems.

This book concentrates on instruments and networks based on RF wireless systems, therefore it is worth revisiting frequency allocation to clarify the operation characteristics of RF networks.

3.6.1 Recent Developments in Frequency Allocation

Frequency allocation is carefully managed by national and international bodies. In the United States, some of the most common bands targeted for wireless single-band or broadband operations are 39 GHz, local multipoint distribution systems (LMDSs), multichannel multipoint distribution systems (MMDSs), Unlicensed National Information Infrastructure (U-NII), 900 MHz ISM, and 2.4 GHz unlicensed bands. This will be explained briefly.

Licensees at 39 GHz provide fixed communication including point-to-point and point-to-multipoint communication. The FCC auctioned and awarded 14 licenses consisting of paired 50 MHz channel blocks in each of the 172 economic areas (EAs) and 3 EA-like areas covering the U.S. territories.

Local multipoint distribution systems offer subscribers a variety of one- and two-way broadband services, such as video programming distribution, teleconferencing, wireless local loop telephony, and high-speed data transmission (e.g., Internet access). LMDSs consist of multicell distribution systems with return path capability within the assigned spectrum. Generally each cell contains a centrally located transmitter (hub), multiple receivers or transceivers, and point-to-point links interconnecting the cell with the center and/or other cells. The FCC offered two licenses, blocks A and B, in each of 493 basic trading areas (BTAs), with a total of 1300 MHz of spectrum per BTA—one for 1150 MHz (block A) and one for 150 MHz (block B).

Multichannel multipoint distribution systems consist of multichannel distribution service, multichannel multipoint distribution service, and instructional television fixed service. Traditionally considered a "wireless cable" band for television channels, MMDS is increasingly being used for implementation of broadband wireless access systems. These broadband wireless systems are intended to provide digital services such as Internet access, WANs, and voice-over IP (VoIP).

The U-NII bands are allocated at the 5.15–5.25 GHz, 5.25–5.35 GHz, and 5.725–5.825 GHz frequencies for use by a new category of unlicensed equipment called U-NII devices. These devices provide short-range, fixed, point-to-point, high-speed wireless digital communication on an unlicensed basis. The 5.15–5.25 GHz portion is intended for indoor use in short-range networking devices. The limit for power is 200 mW in short-range WLAN applications. Devices operating between 5.25 GHz and 5.35 GHz are intended for communication within and between buildings, such as campus-type networks. U-NII devices in the 5.25–5.35 GHz frequency range are subject to a 1 W power limit. The 5.725–5.825 GHz portion of the band is intended for community networking communication devices operating over a range of several kilometers. Point-to-point U-NII devices with up to 1 W transmitter power and directional antennas with up to 23 dBi gain are permitted to operate in this frequency band.

The 900 MHz ISM and 2.4 GHz unlicensed bands support spread spectrum operation on a noninterference unlicensed basis. Operation in this band is authorized under FCC Rule Part 15.247. Spread spectrum systems share these bands on a noninterference basis with systems supporting critical government applications, such as the airborne radiolocation systems.

In the late 1980s the FCC first provided license-free bands under Part 15 of the FCC regulations for low-power spread spectrum devices in the 902–928 MHz, 2400–2483.5 MHz, and 5.725–5.825 MHz bands. In 1997 the FCC extended the unlicensed spectrum to 5.150–5.350 GHz for the purpose of supporting low-power license-free spread spectrum data communication. This allocation was called the U-NII band. This extension is important because it shows the need for allocating more frequencies for wireless systems and networks, and it also indicates the increase in demand. The extension follows the allotment of a much earlier allocation of unlicensed spread spectrum bands by the FCC in the 1980s.

3.6.2   Types of Wireless Networks

The classification of wireless networks depends on factors such as the role of wireless components in the network, the size of the network, and the topology of the network. Modern wireless networks can be divided into three primary categories

- Local area networks—wireless components act as in an ordinary LAN, usually providing connectivity between devices.

- Extended LANs— wireless components are used to extend the range of a LAN beyond the normal distance limitations or normal installation capabilities.

- Mobile networks—individual mobile devices and users communicate using a wireless networking medium, such as radio or cellular telephone systems.

Both wireless LANs and extended LANs involve equipment that a particular organization may own and control. Mobile network applications typically involve a third party supplying the infrastructure necessary for transmission and reception. Most often the company that provides such facilities and services is a communication carrier offering voice and data communication to its customers.

The widespread use of mobile communication has led to the development of many wireless systems and standards that can serve many other types of applications. For example, new generation cellular networks are designed to facilitate high-speed data communication as well as voice transmission. New standards and technologies are being implemented that will allow wireless networks to replace fiber optics or wires between fixed points. Similarly wireless networks have been increasingly used as a replacement for wires in homes, buildings, and office settings through the use of WLANs. For instance, the fast evolving Bluetooth standards may replace the communication cords of many domestic and industrial appliances.

Used primarily within buildings, WLANs and Bluetooth use low power levels and generally do not require a license for spectrum use. These license-free networks provide a dichotomy in the wireless market, since ad hoc data rates are being developed by individual devices operating within buildings. It appears that wireless markets (e.g., in buildings and industry) will become a battleground between licensed and unlicensed services.

3.6.3 Wireless Network Topologies

Wireless networks have the potential to eliminate the need for a visible physical network topology. This is one of the reasons for their rapid growth in popularity. Wireless networking may not need to have a particular topology associated with it; it can be configured to act like a particular topology. Although ad hoc topologies are possible, generally WLANs require a centralized device to control and coordinate communication among its members. In this respect, most WLANs use star topology, since all of the signals travel through a central hub. Readers can find comprehensive information on wireless network topologies in Chapter 4.

3.6.4 Wireless Extended LAN Technologies

Certain kinds of wireless network equipment extend LANs beyond their normal distance limitations. Devices such as bridges and repeaters provide connectivity outside the area of coverage. For example, a typical wireless bridge can connect networks up to 4 km to 5 km.

Such LAN bridges permit the linking of locations within the line of site or outside the line of site when suitably located. LAN bridges may also make it unnecessary to rout dedicated digital communication lines from one site to another through a communication carrier. Spread spectrum radio, infrared, and laser-based equipment used as bridges are readily available on the commercial market. Longer-range wireless bridges are also available, including spread spectrum solutions that work with Ethernet, token ring, and others, ranging over distances of up to 40 km.

As wireless data rates increase and worldwide standards begin to converge, new and wider applications of WLANs are becoming evident. Numerous companies have been exploring a public LAN concept whereby a wireless Internet service provider (WISP) builds a nationwide infrastructure of WLAN access points in selected locations. Commonly used WLANs are based on the IEEE 802.11 standards.

3.6.5 IEEE 802.11 WLAN Standards

The IEEE 802 standards are widely applied in wireless sensor and instrument networks and therefore they will be briefly revisited to highlight important

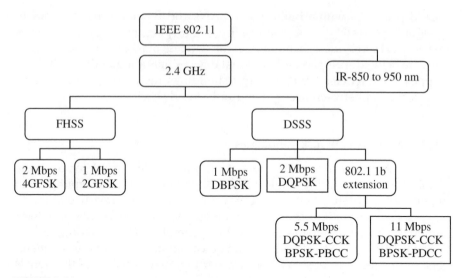

FIGURE 3.14
IEEE 802.11 wireless LAN standards.

points. Figure 3.14 illustrates the evolution of the IEEE 802.11 WLAN standards. In these standards, both frequency hopping and direct sequence approaches are used. Not shown in the figure is the IEEE 802.11a standard, which provides up to 54 Mbps throughput in the 5 GHz band. The DSSS IEEE 802.11b standard has been named Wi-Fi by the Wireless Ethernet Capability Alliance (WECA), a group that promotes adoption of 802.11b DSSS WLAN equipment and interoperability between vendors. IEEE 802.11g is developing Complimentary Code Keying Orthogonal Frequency Division Multiplexing (CCK-OFDM) standards in both the 2.4 GHz (802.11b) and 5 GHz (802.1a) bands, and supports roaming capabilities and dual band use for public WLANs.

Figure 3.15 illustrates WLAN channels that are specified in the IEEE 802.11b standard in the 2400–2483.5 MHz band. All WLANs are manufactured to operate on any one of the specified channels and are assigned to a particular channel by the network operator when the WLAN system is first installed. All wireless systems must be designed with knowledge of the interference and propagation environment; prudent WLAN deployment dictates that the placement of transmitters and their frequency assignment must be done systematically to minimize impact. Even though WLANs are designed to work in an interference-rich environment, specific placement of access points can provide significant improvement in cost and end-user data throughput in heavily loaded systems. The channelization scheme used by the network operator becomes very important for a high-density WLAN installation to avoid interference and maintain good performance.

There is a range of software available (e.g., SitePlanner from Wireless Valley) for the deployment of WLANs. With the use of such software a

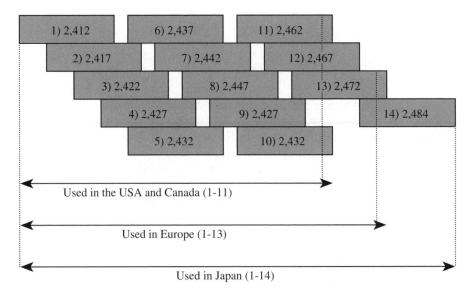

FIGURE 3.15
A 2400 MHz channelization scheme for IEEE 802.11.

WLAN can be set up quickly with a minimum need for trial and error. These programs use the blueprints of the buildings where the WLAN will be installed to predict the characteristics of wireless operations, including interference, the best location for access points, future expansions, etc. There are other WLAN software products (e.g., LANfielder and Sitespy) for collecting and using the performance of in-the-field network measurements for network validation and to obtain optimal WLAN designs.

3.6.6 The HiperLAN Standard

In Europe, in the mid 1990s, the HiperLAN standard was developed to provide similar capability as the IEEE 802.11 standards. HiperLAN is primarily used in European countries. There are two specifications: HiperLAN/1 and HiperLAN/2. Both have been adopted by the European Telecommunications Standards Institute (ETSI).

HiperLAN/1 provides communication at up to 20 Mbps in the 5 GHz range of the RF spectrum. HiperLAN/2 operates at up to 54 Mbps in the same RF band. HiperLAN/2 is compatible with third-generation WLAN systems for sending and receiving data, images, and voice communication. HiperLAN/2 has the potential and is intended for implementation worldwide in conjunction with similar systems in the 5 GHz RF band.

HiperLAN is primarily intended to provide individual wireless LANs for computer communication, but at the moment it is being used for many other applications. HiperLAN provides asynchronous user data rates of between 1 Mbps and 20 Mbps, as well as time-bounded messaging at rates of 64 kbps to

2.048 Mbps. It was designed to operate in moving environments at speeds of up to 35 km/h, and typically provides 20 Mbps throughput at a range of 50 m.

3.6.7 Bluetooth

There is a vast consumer market for wireless technology. Wireless technology provides both convenience and flexibility. It also allows collaborative communication between devices and appliances as well as central controllers.

Bluetooth is a standard for short-range, low-power, low-cost wireless communication. It is an open standard that has been embraced by thousands of manufacturers of electrical appliances (www.bluetooth.org). Some of the common appliances that use Bluetooth are cell phones, PCs, mice, keyboards, joysticks, cameras, printers, audio handsets, speakers, and musical instruments. It provides an ad hoc approach for enabling various devices to communicate with one another with a nominal distance of 10 m.

Bluetooth complies with FCC Part 15.247 rules and has a restricted maximum allowable peak power of 1 W. The FCC also regulates that at least 75 of the 79 channels should be used in a pseudorandom manner and a device cannot operate for more than 0.4 s on a given channel within any 30 s period. These restrictions also apply to 802.11b/g devices and they are deemed necessary to minimize the effect of interference.

Bluetooth operates in the unlicensed 2.4 GHz ISM band. In most countries, there are 79 channels available for use; however, some countries only allow 23 channels. The nominal bandwidth for each channel is 1 MHz. Bluetooth uses the frequency hopping technique at a hopping rate of 1600 times per second with a residence time of 625 μs. Transmissions are performed in 625 μs slots with a single packet transmitted over a single slot. For long data transmissions, a particular user may occupy multiple slots using the same transmission frequency, thus slowing the instantaneous hoping rate to less than 1600 hops per second. In an inquiry or page mode, it hops 3200 times per second with a residence time of 312.5 μs. The frequency hopping scheme of each Bluetooth user is determined from a cyclic code with a length of $10^{27}-1$, and each user has a channel symbol rate of 1 Mbps using Gaussian frequency shift keying (GFSK) modulation.

The Bluetooth specification uses TDD and TDMA. The time slot for a single packet is 625 μs. All communication between devices take place between the master and slave using TDD. The master polls each active slave to see if they have data to transmit. If the slave has data, it must send it immediately after it has been polled. The master transmits only in even numbered time slots, while slaves transmit only on odd numbered time slots. In each time slot, a different frequency channel is used for the hopping sequence.

At the baseband layer, a packet consists of an access code, a header, and a payload. The access code is usually 72 bits long and contains device addresses. The header is 18 bits long and includes the active member address. The header also contains link control data encoded with forward error cor-

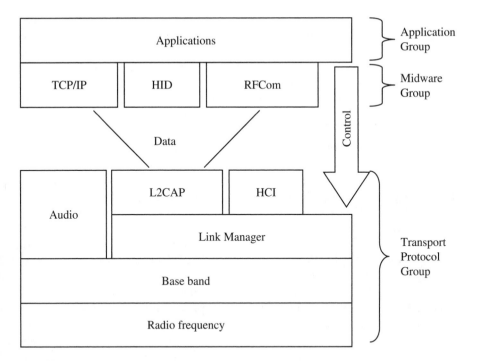

FIGURE 3.16
Bluetooth protocol groups.

rection (FEC) code, with a rate for high reliability. The payload can contain from 0 to 2745 bits. For synchronous connection oriented (SCO) systems, packets must be exactly one time slot in length. For asynchronous connection less (ACL) links, packets may be one, three, or five time slots in length.

The Bluetooth specification divides the Bluetooth protocol stack into three logical groups: transport, midware, and application. The Bluetooth core protocol groups are illustrated in Figure 3.16.

Transport group protocols (equivalent to data link and physical layers of the OSI model) allow Bluetooth devices to locate each other, and manage the physical and logical links with higher layer protocols and applications. Radio, baseband, link manager, local link control, and adaptation (L2CAP) layers, and the host controller interface (HCI) are included in this group.

The radio layer is concerned with the design of transceivers. The baseband layer defines how Bluetooth devices search for other devices and how they establish connections. The master and slave arrangements, typical of Bluetooth, are made at this level. RF communication is realized by frequency hopping sequences and TDD. Packet transmission and retransmission, packet processing procedures and strategies for error detection and correction, signal scrambling, and encryption are arranged in this layer. The link manager layer implements the link manager protocol (LMP), which manages the properties of air interface links between devices. The bandwidth allocation for general data and the bandwidth reservation for audio traffic, authentication, trust relations between devices, encryption of data, and internal and RF power use are under the control of the LMP. The L2CAP layer provides an interface between higher level and lower level transport protocols. Link manager is also responsible for multiplexing of higher level protocols, packet segmentation and reassembly, and maintenance of services between devices. The HCI enables interoperability and defines a standard interface for upper level applications to access the lower layers of the stack. All the protocols in the group are required to support communication between Bluetooth devices.

The midware protocol group includes third-party and industry standard protocols as well as Bluetooth Special Interest Group (SIG)-developed protocols. These protocols allow already available and new applications to operate over Bluetooth links. Industry standard protocols include the point-to-point protocol (PPP), internet protocol (IP), transmission control protocol (TCP), wireless application protocol (WAP), and object exchange (OBEX) protocols adapted from the IrDA.

The application group depends on the application in hand. The Bluetooth SIG-developed protocols include a serial port emulator (RFCOMM) that enables legacy applications to operate seamlessly over Bluetooth transport protocols. Other protocols developed by the SIG are the service discovery protocol (SDP), which allows devices to obtain information about each others' available services, and the packet-based telephony control signaling (TCS) protocol, which manages telephony operations. The reuse of existing protocols and seamless interfacing for existing applications were given the highest priority during the development of Bluetooth specifications.

The Bluetooth standard is designed to support communication in very high interference environments and relies on a number of FEC and coding techniques. An ARQ scheme supports a raw channel bit error rate of about 10^{-3}. The nominal range of transmission is about 10 m. This range can be increased to 100 m by using amplifiers.

Different countries have allocated various channels for Bluetooth operations. In the United States and most European countries, the frequency hopping spread spectrum (FHSS) 2.4 GHz ISM band is available. The wireless personal area network (WPAN), based on Bluetooth specifications, is now an IEEE standard, 802.15. IEEE 802.15 aims to interconnect pocket PCs, personal digital assistances (PDAs), cell phones, and many other appliances.

3.6.8 Industrial Sensor Buses and Networks

Industrial instruments and sensor systems are built on the general communication buses and application-specific sensor buses that are available for factory automation, process control, home automation, and other applications. However, modern sensors, particularly smart sensors, can be directly connected to computers and networks. These connections can be realized by using point-to-point and multiplexing techniques when there is more than one sensor involved in the system. In conventional systems, point-to-point and multiplexed connections require substantial amounts of wiring, and are bulky and costly to build and maintain. However, with the emergence of wireless technology and networks, manufacturers and users alike are keen to apply wireless techniques in industrial environments. As a result, many wireless industrial sensors, control networks, and hybrid networks that integrate wireless and wired media are appearing in the marketplace.

The sensor buses are basic devices for connecting sensors and actuators to transfer information to and from host computers. In these buses, the sensor data are primarily passed straight through the system and are seldom processed. The device-level control networks are higher level buses, and format and perform some basic processing of the data at each network node before passing it to the host computer. The fieldbus networks incorporate the user layer that resides on top of the application layer and can perform control procedures in field devices as well as in the controller. The fieldbus differs from device-level networks in its ability to provide intrinsic safety. Device-level networks can be used in conjunction with fieldbusses to handle discrete automation functions because of their ability to handle small packets of information at high speeds. Device-level networks are suitable for devices with limited data and for dedicated applications such as machine control and condition monitoring.

In industrial applications, the idea of a sensor bus was originally evolved from wired sensor-driven control systems. In a sensor bus, a host computer interfaces with multiple nodes over a bidirectional digital data and control bus, as illustrated in Figure 3.17. Each node of the sensor bus consists of one or more sensors or actuators.

A sensor bus or control network consists of a communication bus or network to which a number of device nodes are connected to a computer. The computer communicates with the devices using a standard protocol to implement sensing and control, or monitoring applications. Communication among the nodes can be by either a peer-to-peer or master-slave arrange-

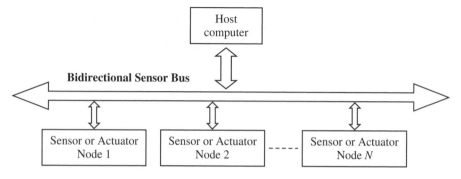

FIGURE 3.17
A basic sensor bus.

ment. In either case, intelligence in the nodes, in terms of computational capability, allows the distribution of processing loads. If smart sensors are used, they can perform local data analysis and conversion, and report only important information.

One of the basic sensor buses is known as the Michigan Parallel Standard (MPS) bus. This bus contains 16 lines—8 bidirectional data lines, 4 control lines, 1 parity line, and 3 power lines. Another basic sensor bus is the Michigan Serial Standard (MSS) bus, which has only four lines. The MSS bus is similar to the RS-232 and RS-499 standards for two-wire half-duplex operation, but it uses a single data line and its protocols are custom designed for sensing systems with central controls. This bus can accommodate up to 256 nodes.

In parallel to MPS and MSS, Delft University of Technology developed the integrated smart sensor (IS^2) bus using four lines. This bus uses a single master control for data transfer to and from slave devices or nodes. Besides the two lines used for power, one of the lines is used for bidirectional serial data communication and the other is designated as the bus clock.

Some examples of the sensor buses, device control networks, and field network protocols that can be integrated together to form complete industrial sensor networks are listed below:

- ARCNET—Attached Resource Computer Network; operates with a maximum data rate of 2.5 Mbps.

- ASI—Actuator Sensor Interface; a low-cost, bit-level system designed to handle four bits per message for binary devices arranged in a master-slave structure. It operates over distances of up to 100 m in factories and process control environments.

- BACnet—Building Automation Control Network; supports networking operations in buildings in conjunction with other protocols such as ARCNET, Ethernet, RS-485, and LonWorks.

- BITBUS—also known as IEEE 1118; features multitasking operations with a master-slave structure using an RS-485 serial link. The master can accommodate up to 249 devices.

- CAN—Controller Area Network, also known as ISO-11898; uses twisted pair, fiber-optic, or coaxial cables or RF media for communication and offers selectable baud rates up to 1 Mbps. CAN busses are used extensively in cars, trucks, and other transport vehicles.

- CEbus—Consumer Electronic Bus; supports a data rate of 10 kbps and can access a maximum of 61,000 nodes. It supports coaxial cables, infrared, RF, twisted pair, and power line media.

- DeviceNet—built on top of CAN and features the object-oriented software suitable primarily for industrial control systems. It uses four-wire shielded cable and supports up to 64 nodes per network at speeds of 500 kbps at 100 m and 125 kbps at 500 m.

- Ethernet—a 10 Mbps LAN used primarily for office and enterprise automation. It is becoming a very strong contender for control and field networks for the process and manufacturing industries because of its speed, real-time control features, and readily available cabling infrastructure. Much higher Ethernet communication speeds are available up to 1 Gbps.

- FireWire—also known as IEEE-1394; a high-speed interface for connecting computers and peripherals. FireWire supports up to 63 devices with a maximum data rate of 400 Mbps and is used to connect computers and consumer electronic devices such as digital cameras and music systems.

- Foundation Fieldbus—formed by merging the components of specifications by WorldFib and Profibus. Extensive information on Foundation Fieldbus is available in Chapter 1.

- GPIB—General Purpose Interface Bus, also known as IEEE 488; used for data acquisition systems with limited node capabilities in laboratories and industrial instrument systems.

- HART—Highway Addressable Remote Transducer; provides two-way digital communication atop traditional 4 mA to 20 mA loops.

- Interbus S—a fast sensor/actuator data ring-type bus. For digital data flow, it utilizes RS-422 technology, but can handle analog I/O. It supports 256 drops per network and up to 4096 digital I/Os.

- ISA SP50—based on the traditional 4 mA to 20 mA standard for the process industry.

- J1850—a standard for passenger cars covering midspeed data rates optimized at 10.4 kbps and 41.6 kbps.

- LonWorks—Local Operating Network; uses custom neuron chips implementing the ISO/OSI seven-layer stack protocol. It supports media such as twisted pair, coaxial cable, fiber optics, RF, infrared (IR), and power line with data rates up to 1.25 Mbps for distances of 500 m and 78 kbps at 2000 m.

- Profibus—Process Field Bus, also known as DIN-19245; consists of four parts. Part 1 and Part 2 are designated as Profibus-FMS and cover automation in general. Part 3, Profibus-DP, is a faster system for factory automation. Part 4, Profibus-PA, is for process control.

- SDS—Smart Distributed System; a CAN-based system using four-wire cable. It supports up to 128 nodes at speeds of 1.25 Mbps for interfacing with programmable logic controllers (PLCs) and PCs in industrial control applications.

- SERCOS—Serial, real-time Communication System; uses a fiber-optic ring configuration and can accommodate up to 24 nodes on the ring.

- SERPLEX—an application-specific integrated circuit (ASIC)-based multiplexing control bus that offers both peer-to-peer and master-slave communication.

- WorldFIP—Factory Information Protocol; based on three OSI reference model layers: 1, 2, and 7. It is supported by chip-level and board-level products.

Many of these industrial networks support wireless sensors and instruments.

3.7   Network, Data, and Information Management

Networks constitute many interconnected devices as well as communication infrastructures between these devices. To support these operations, relevant network management software must be implemented and management plans must be laid out, requiring careful creation of infrastructure for servers, routers, printers, switches, firewalls, etc.

3.7.1   Network Management

Network management entails supervision of the information technology system, including purchasing of equipment and software, physically distributing and installing it, configuring it, maintaining it, providing enhancements and service updates, setting up problem-handling processes, and determining whether objectives are being met. It is a well-established, but still growing field, particularly in the case of wireless networks. Systems management can also refer to the class of software that automates various processes associated with overseeing both the hardware and software, including activities such as asset tracking, software versioning, performance measurement, availability monitoring, data backup, capacity planning, and disaster recovery.

All digital systems use software, which are programs of instructions. To run a program, a digital device loads it from the storage device into primary memory. The central processing unit (CPU) then executes the instructions. Programs are generated by languages such as machine codes, assembly language, or high-level languages. There are many high-level languages: Fortran, Cobol, Basic, Algol, Pascal, C, C++, Ada, Java, and so on. There are different levels of software for network management, including operating systems, data management software, and information management software.

The operating system (OS) is an integrated set of programs used to manage resources and overall operations of digital systems, computers, and networks. It permits the system to supervise its own operation by automatically calling in application and service programs, and interacting with users and other connected devices. For computers, for example, there are many different types of operating systems, such as MS-DOS, OS/2, Windows, and Mac OS for personal computers, and Unix and Linux for larger machines.

Data management is an essential part of network management. The main elements of data management are the generation, transferring, storage, and application of data. The generation of data relevant to instruments and instrumentation is explained extensively in Chapter 1. Data transfer involves appropriate communication systems and associated hardware and software support. Data can be transferred by circuit switching, as in Public Access Branch Exchange (PABX) systems, or by message switching, packet switching, cell switching. Packet switching is a popular method; it splits the messages into short packets and sends them to the destination in an efficient way. Cell switching makes use of small, fixed-length cells that can be transferred between the source and the sink very quickly. Data storage requires careful attention, particularly in identifying those data that are relevant and necessary in the long term. Data applications in instrumentation systems depend on the system layout (e.g., process flow in a production environment).

Information management varies from one system to another. Information is a pattern of symbols that have some real-world meaning. It is data that are arranged and convey some useful message. For example, in industrial applications, information concerning machine health and safety may be of prime importance. Information management is concerned with the generation, storage, and use of information in a timely, secure, accessible, and complete manner.

There are three main requirements for information management: an information system, which can be defined by the applications required and the databases that store and provide information; information technology, which includes the hardware, software, and communication infrastructure of an information system; and information management, which requires policies, methods, performance measurements, process and data models, applications, and communication of the information within an organization.

Policies in information management determine the guiding principles, regulations, and directions for the information managers in an organization. Methods lay out the principles of information system technology. Process and data models provide the guidelines, documentation, and task execution for the interpretation and execution of information. Applications of information maintain the desired functionality and interoperability between applications. Communication is important to support efficiency and decision making.

3.8 Conclusion

This chapter concentrated on both wired and wireless networks. The technology for wired networks has been in use for many years, but wireless networks in sensors and instruments are relatively new. However, the knowledge and experience gained from wired technologies is being applied directly to wireless operations. Newly emerging wireless technologies such as IEEE 802 and Bluetooth were discussed extensively and revisited when necessary. Protocols and standards that are relevant to wireless networks were explained in detail. The importance of security issues in wireless networks were stressed and various methods discussed.

4

Wireless Instrument and Sensor Networks

In this chapter, all the information provided in Chapter 1 through Chapter 3 are brought together. After having read this chapter, readers will be able to configure various forms of wireless instruments as well as complex networks. The reader will be able to select appropriate components for the application and will be able to appreciate important issues in the wireless networking of sensors and instruments.

In the previous chapters, it was pointed out that the sudden growth in communication technology has prompted expansion of the wireless industry by orders of magnitude and at the same time has opened up many new application areas. The growth in wireless technology is largely supported by improvements in digital and radio frequency (RF) circuit fabrication methods, advances in signal processing theory and applications, and the emergence of new large-scale low-power integrated circuits (ICs) and other supporting devices. In particular, new IC technologies have resulted in smaller, cheaper, and more reliable radio equipment. In parallel with the expansion and diversity of wireless devices, consumer acceptance and need has fueled the widespread expansion of wireless communication systems.

In addition to the rapid development of modern instruments, advances in IC technology, microelectromechanical systems (MEMS), intelligent sensors, and wireless communication technology have made large-scale and inexpensive sensor networks possible. The distinction between an instrument network and a sensor network is that, although they use the same basic elements, instruments have additional components such as displays, control buttons. However, a new trend is to use sensing elements that can directly communicate with external devices. This offers many advantages over traditional instrumentation methods, including large-scale and dense deployment for larger systems with higher resolution, better fault tolerance and more robust systems, configuring of ad hoc networks, and cost savings. These systems can be used in application areas such as the military, habitat monitoring, environmental observations, and many others.

Modern wireless sensors and instruments and their networks can be produced in a number of different ways, such as

- Embedding the wireless communication system into the sensor or instrument.

- Adding wireless modules to existing sensors or instruments. This technique uses off-the-shelf transceivers or application-specific transceivers integrated with the sensor or instrument.

- Connecting modems such as wireless RS-232, wireless universal serial bus (USB), etc.

- Using bridges, repeaters, gateways, and data loggers.

In this chapter, the construction of wireless sensors and instruments using these methods and the network architectures are explained and examples provided. Instrument communication protocols are revisited and current technologies applied in wireless instruments and sensors are explained. Applications of Bluetooth and IEEE 802 technologies are introduced with some examples and industrial wireless sensor and instrument networks are discussed.

One advantage of wireless networks is that they offer communication alternatives among the devices involved. The communication can take place as point-to-point or point-to-multipoint. This gives considerable flexibility in network configurations, and communication algorithms can be tailored to improve reliability and adaptability.

4.1   Wireless Sensor Architecture and Network Design

Semiconductor IC technology resulting in the proliferation of intelligent sensors and microsensors and progress in digital RF communication systems supported by the relevant software are adding remarkable features to instrumentation systems and opening up many potential application areas. Wireless technology in instrument networks is not new. Industry has been using wireless instrument networks for many years via supervisory control and data acquisition (SCADA) systems or the like. Wireless sensor networks, on the other hand, are relatively new and have primarily been driven by mostly environmental and military applications such as battlefield surveillance, tracking, and environment monitoring. Today, wireless sensor and instrument networks are used in many applications, including habitat monitoring, environmental observation and monitoring, health care, and many others.

The most fundamental choice to be made in the deployment of wireless sensor networks is selection of the transmission frequency. One option is the use of licensed bands in the very high frequency (VHF) and ultra high frequency (UHF) regions. The licensed bands require an application procedure through the U.S. Federal Communications Commission (FCC), or leasing of transmission facilities and/or services from independent providers such as private industrial radios, shared trunked systems, cellular digital modems, or satellite-based systems. Another option is the use of unlicensed systems, but these have restrictions on the maximum power that can be transmitted in the industrial, scientific, and medical (ISM) bands. The technology supporting ISM bands is developing quickly, providing many different application possibilities for end users. It is also encouraging vendors to supply a diverse range of products, which in turn is prompting major changes in wireless technology. The ISM bands used in different regions of the world include 13.56 MHz, 27.55 MHz, 303 MHz, 315 MHz, 404 MHz, 433 MHz, 868 MHz (Europe), 915 MHz (North America), 2.45 GHz, 5.2 GHz, 5.3 GHz, and 5.7 GHz (North America).

The ISM bands offer implementation flexibility, but users should be concerned with the possibility of interference in those bands. However, spread spectrum technology and cryptography allow users to achieve satisfactory reliability and efficiency.

4.1.1 Wireless Sensors and Transducers

Modern wireless sensors are configured as either embedded devices or as modular devices using add-on components. In embedded sensors, the wireless communication components are integrated with the sensor in the same chip. In modular devices, RF modules such as transceivers are added externally on the existing sensors.

Both methods are based on digital IC technology supported by other digital components and appropriate input/output (I/O) systems. There are many types of sensors commonly used in wireless sensor operations, including IC sensors, MEMS, web sensors, intelligent sensors, wireless sensors, and application-specific integrated circuits (ASICs). These sensors and their wireless operations will be explained next.

Integrated circuit sensors are the backbone of wireless devices. Many of these devices contain the sensor, signal processors, and logic circuits in a single chip. They are available as accelerometers, biosensors, chemical sensors, optical sensors, magnetic sensors, pressure sensors, custom silicon microstructures, etc. In many applications, IC sensors are equipped with digital devices and radio transceivers that enable wireless operation.

Most wireless sensors are made from silicon. The semiconducting properties of silicon have made it a basic building block of the electronics industry. Silicon also has excellent physical properties that make it an ideal material for mechanical devices. It has a tensile strength greater than steel and is almost perfectly elastic, making it a suitable material for use in MEMS products. It is free of hysteresis, and its crystalline structure is well suited to the fabrication of miniature precision products. These silicon micromachined products have several advantages over their conventionally manufactured counterparts: they are generally much smaller, performance is higher due to precise dimensional control in fabrication, and costs are lower because thousands of units can be produced at one time.

A two-axis digital inclinometer from Crossbow Technology Inc. (CXTILT02E; http://www.xbow.com) is a good example of an IC sensor. This inclinometer measures the tilt angle of an object with respect to horizontal in a static environment. To measure tilt, also called roll and pitch, the sensor makes use of two micromachined accelerometers, one oriented along the *x*-axis and one along the *y*-axis. The sensor has an embedded microcontroller and analog-to-digital (A/D) converter. The resolution and settling time of the sensor are programmable, allowing it to be customized for any application. It has an RS-232 interface and an on-board temperature sensor block to internally compensate for temperature-induced drift.

Along with system-on-a-chip wireless capabilities, MEMS are becoming commonly available. While the components of MEMS have been around for some time (e.g., microsensors, microactuators, microelectronics, and microstructures), we are now able to combine them in a reliable and cost-effective manner.

With today's MEMS technology, combining the radio transceiver, sensors, and circuitry is still in its infancy. MEMS technology tends to operate at low frequencies, and sensor response may be sensitive to low-frequency interference. In the case of accelerometers as the sensing element, sensitivity to environmental conditions such as shock becomes problematic.

An example of a wireless MEMS sensor is an accelerometer from Micro-Strain (http://www.microstrain.com/) illustrated in Figure 4.1. This sensor has a triaxial accelerometer that operates over 2.4GHz radiofrequency, direct sequence spectrum, and uses the new IEEE 802.15.4 standard. It transmits data continually at 1 kHz sweep rates for a preprogrammed time period. The continuous wireless transmission mode allows for real-time data acquisition and display from a single multichannel sensor node. This accelerometer can be used to monitor tilting, vibrating, and rotating machinery, equipment, and structures. Multiples of these sensors can be networked in a wide area, distributed over many different machines and locations. This device is supported by a PC-based software package for Windows 98/2000/XP machines. The software provides an interface to all functions including data logging, erasing, streaming, configuring, graphing, data file saving, and wizards. Wizards include accelerometer calibration, automatic offset balancing, gain adjustment, and custom formulas supporting user-specified data scaling.

Viable MEMS are being produced with radio components. Researchers at Georgia Tech, the Defense Advanced Research Projects Agency (DARPA), Carnegie Mellon University, the University of California at Berkeley, and others have done extensive research on MEMS sensors in conjunction with radios. The announcement by MEMSCAP in May 2000 of a component library of MEMS inductors and variable capacitors heralds a new generation of radio design.

Web sensors are also becoming very popular devices because they enable Web connections without the need for any other intermediate support. They are particularly useful where remote data acquisition is required. Internet-accessible sensors have built-in signal conditioning and a mini-Web address

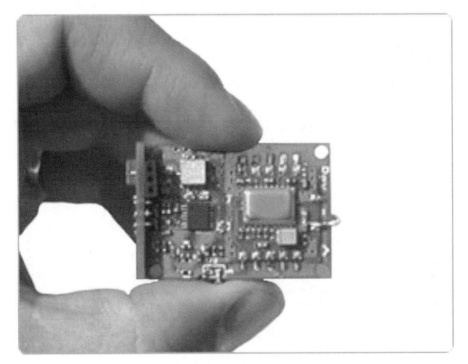

FIGURE 4.1
An MEMS-based wireless accelerometer (Courtesy of Microstrain, Inc., http://www.micros-train.com/).

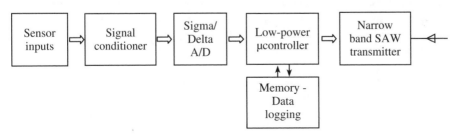

FIGURE 4.2
A wireless Web sensor.

so users are able to access the sensor directly or via a Web site. A typical arrangement of a wireless Web sensor node is illustrated in Figure 4.2.

In recent years, IEEE 1451-compatible interfaces between Internet and Ethernet serial port Web sensors have been developed. These sensors have direct Internet addresses. The interface is realized in IEEE 451 network capable application processors (NCAPs). The NCAP connects the Internet through the Ethernet. The NCAP is a communication protocol that allows receiving and sending of information using standard transmission control protocol/Internet protocol (TCP/IP) ports. Through the NCAP protocol, sensor data can be formatted to be sent or received from serial ports. The serial ports can be RS-232, RS-485, TII, Microlan 1-wire, Esbus, or the inter-IC bus (I^2C).

The EM01 Web sensor from Esensor Inc. (http://eesensors.com) is an example of a web sensor. The EM01 is able to monitor temperature, humidity, and illumination. The main parts of the sensor consist of a base unit together with two communication ports (Ethernet and Esbus). The Esbus combines a networked version of the serial peripheral interface (SPI) bus and RS-232. The Ethernet is the route to the Internet. Within the base unit are microcomputers that handle the Internet protocol, the communication ports, and in some cases, the sensor and sensor signal conditioning. A Web sensor has no switches, keyboard, or (required) display because it functions as a sensor, not as a computer. Additional sensors can be connected to the base unit by an Esbus (networked), RS-232, or IEEE 1451 port, depending on the model.

Intelligent sensors are advanced microsystems that include a sensor, interface circuits, and pattern recognition in the same chip. They are used extensively in instruments and instrumentation. Some of these sensors are manufactured with the Neural Network and others with sophisticated intelligence techniques such as Fuzzy Logic circuits and software on board the chip. There are many different types of intelligent sensors, including neural processors, vision systems, industrial systems, and intelligent parallel processors.

An example of an intelligent wireless sensor is one from the Oak Ridge National Laboratory as shown in Figure 4.3a and b. This is a second-generation integrated wireless sensor that has both on-chip and off-chip sensor capabilities. The optical data port provides a wireless interface for changing gain, update rate, and other parameters using a TV remote control instead of jumpers. This is part of its migration to a fully bidirectional telesensor chip.

Another example is an intelligent sensor is one that combines five functions in one device: sensing with local analog or digital readout, transmission of a 4 mA to 20 mA process signal, set-point control, switching by 3 A relay or transistor output, and plug-and-play connection to a fieldbus. Units connect directly to valves and can interface to programmable logic controllers (PLCs) or operate without PLCs, with parameters set on the unit or by a PLC. Calibration can be performed without operating the system.

There are many other examples of intelligent industrial sensors. For instance, a chemical vapor sensor uses an artificial neural network (ANN) to learn to identify various chemicals, gases, and acids after being trained. It recognizes airborne volatile organic compounds, alcohols, ethers, halocarbons, carbon monoxide, carbon dioxide, warfare stimulants, etc. Some sensors use recurrent neural network (RNN) learning for chemical classification.

Wireless sensors represent complete miniaturized systems that contain sensing elements, signal processing circuits, and wireless communication components in the same chip supported by some external components (Figure 4.4). Several miniaturization techniques are available for wireless sensors, including system-on-chips, MEMS, and ASICs. In a wireless sensor,

Advanced Wireless Telesensor Chip

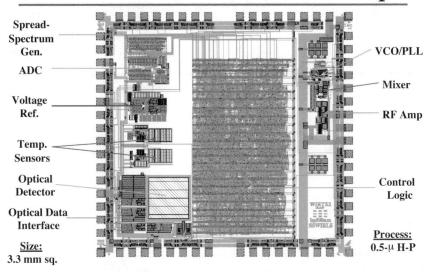

Spread-
Spectrum
Gen.

ADC

Voltage
Ref.

Temp.
Sensors

Optical
Detector

Optical Data
Interface

Size:
3.3 mm sq.

VCO/PLL

Mixer

RF Amp

Control
Logic

Process:
0.5-µ H-P

Intelligent Wireless Sensors & Systems

INTELLIGENT WIRELESS SENSOR
2nd multiple-sensor proof-of-principle device

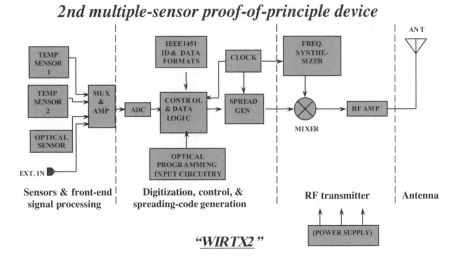

"WIRTX2"

Intelligent Wireless Sensors & Systems

FIGURE 4.3
An intelligent wireless sensor design. (Courtesy of Oak Ridge National Laboratory)

FIGURE 4.4
A range of wireless sensors (Courtesy of Crossbow Technology, http://www.xbow.com/).

there are five main components that need to be integrated for a complete system: the sensor, signal processing circuitry, radio, battery, and package. The radio, sensors, and signal processing circuitry can be reduced in size through hybrid circuits, MEMS, or mixed signal ASIC design. However, the power supply components must be dealt with separately.

An example of a wireless sensor is from Crossbow Technology Inc. (http://www.xbow.com/). This company offers a range of products that support each other, such as MICA2 Mote processor/radio boards, base stations, and sensor boards. The MICAz kit includes eight 802.15.4/ZigBee-compliant motes, sensor and data acquisition boards, and two different gateway/interface boards. Other kits include eight motes, sensor and data acquisition boards, and a base station. Wireless kits are supported by end-to-end software. XMesh is software that supports low-power and self-forming reliable networking stacks that run on each mote. XServe is the server software that manages data logging and forwarding of mote network data. MOTE-VIEW is the client software for monitoring, visualization, and network management. In addition to core software components, a number of software tools are provided for system development and debugging of wireless sensor network applications. Additional software includes Surge Network Viewer, for displaying real-time mesh network performance analysis.

Hybrid circuits are designed using thin film hybrid technology or chip-on-board technology and are used to significantly reduce the size of wireless sensors. Standard printed circuit boards with surface-mounted components can be reduced in size by using chip dice, subminiature components, and thin film hybrid technology. The functional block diagram of a hybrid circuit wireless system is illustrated in Figure 4.5.

Combining sensor components with radio circuitry in a thin film hybrid is not as simple and straightforward as it seems. As packages shrink, interference problems between the internal components increase, thus requiring

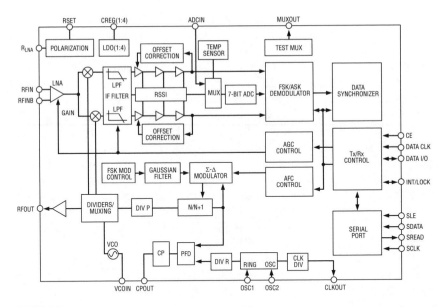

FIGURE 4.5
A functional block diagram of a typical hybrid transceiver (Courtesy of Analog Devices, http://
/www.analog.com/).

more complex circuitry for compensation and component isolation. The
increased complexity of the circuits results in more dissipated power. Packaging a radio in such a small space with microcontrollers and analog accelerometers involves careful design and analysis.

A typical example of a hybrid transceiver is the ADF7020 from Analog
Devices (http://www.analog.com). This device is a low-power, intermediate
frequency (IF) transceiver that operates in the license-free ISM bands at 433
MHz, 868 MHz, and 915 MHz. It is suitable for applications that meet either
the European ETSI EN-300-220 or FCC Part 15.247 and Part 15.249 regulatory
standards. It operates from a 2.3 V to 3.6 V power supply with programmable
output power from –16 dBm to +13 dBm in 0.3 dBm steps. Its receiver
sensitivity is –117.5 dBm at 1 kbit/s in frequency shift keying (FSK) mode
or –110.5 dBm at 9.6 kbit/s. Power consumption is 20 mA in receive mode
and 30 mA in transmit mode (+10 dBm output). Other features include an
on-chip voltage controlled oscillator (VCO), fractional-N phase-locked loop
(PLL), on-chip seven-bit A/D converter, digital received signal strength indicator (RSSI), temperature sensor, and a patent pending fully automatic frequency control loop. This allows the device to be used with lower tolerance
crystals. Its leakage current is less than 1 µA in power-down mode.

Application-specific integrated circuits offer largely hardware solutions as
opposed to purely software (programmable processor) solutions. Historically, developing mixed-signal ASICs is cost prohibitive, and the devices are
sensitive to layout. Only when dealing with volumes in the millions are
ASICs cost effective. Mixed-signal ASICs require external components to
complete the circuits, but this gives flexibility in design and construction.

There are many very large scale integration (VLSI) manufacturing companies, such as Nordic Semiconductor (http://www.nvlsi.no/), that concentrate on a range of devices including RF/mixed-mode devices. Such companies offer many VLSI products, such as standard components for wireless communication, high-performance data converters, and custom design and manufacturing services. Nordic Semiconductor offers 2.4 GHz transceiver, transmitter, and receiver families, such as the nRF24XX. This device is a low-current, low-voltage, 2.4 GHz, Gaussian frequency shift keying (GFSK) single-chip transceiver produced in a 24-pin package. The transceiver, manufactured in a 0.18 μm complementary metal oxide semiconductor (CMOS) process, includes all inductors and filters in a 5 mm × 5 mm package. Two external components needed to make a complete RF system are a crystal oscillator and a resistor. Maximum current consumption of the transceiver is 18 mA in receive mode and 10.5 mA in transmit mode at –5 dBm. Configuration of the nRF2401A transceiver is done via a standard serial interface. It has a data rate of 0 Mbps to 1 Mbps and supports frequency hopping with a channel switching time of less than 200 μs. The power supply range is from 1.9 V to 3.6 V. Such transceivers are used in many applications, including wireless game controllers, PC peripherals, wireless headsets, sports and leisure equipment, toys, radio frequency identification (RFID), remote control and industrial applications, and interactive equipment.

4.1.2   Architecture of Wireless Sensor Networks

In the majority of applications, each sensor must be individually wired to some data monitoring device. Sensor wiring is a tedious, time-consuming, problematic, and manual task demanding high-cost manpower. For these reasons, many companies have developed wireless solutions to avoid the problems associated with wiring sensors. However, there are a number of issues that must be addressed in wireless network architecture. Some these issues are

- Communication must be established at the physical layer and wireless communication protocols must be implemented for data monitoring, encoding, and transmitting.
- The system architecture must provide precise control over radio transmissions. This is particularly important in communication and data collection applications.
- Algorithms must allow the establishment of networks in an efficient and error-free manner.

Wireless networks can be achieved by point-to-point or point-to-multipoint communication methods. In large networks, where there are many wireless sensors operating at the same time, networking requires intelligent sensors and special programming algorithms so that sensors obey some form of a hierarchical structure. The development of networked sensors is progressing in several directions. These are

- Developing low-cost intelligent and programmable sensors backed up by low-cost signal processing hardware and software. This can lead to sophisticated detection methods, identification and tracking functions, and collaboration with other devices in the network.
- Providing long operational life of sensors that are energy efficient and small, housed in smaller and lighter packages.
- Using reduced bandwidths and spread spectrum techniques for wireless information flow between sensors. This also provides somewhat secure data flow and immunization against electromagnetic noise or deliberate interference caused by third parties.
- Minimizing the total sensor network cost and maximizing the sensor network lifetime.
- Providing autonomous sensor management (self-management) techniques and algorithms, such as automatic node selection and self-calibration.

Since the wireless networking of sensors is relatively new technology, the majority of developments have taken place in application-specific fields. Therefore most applications are not quite ready for widespread use in the real world. However, the applications share some common characteristics, such as raw sensor data transmission over wireless connections, centralized data processing, routing schemes, data transport designs, and so on. With progress in sensor fabrication techniques, research on sensor networks, and increasing multidiscipline cooperation, real-world sensor network applications are becoming commonplace.

The future of wireless sensors will probably be shaped by how they interface into the network to share information. In this respect, standards such as IEEE 1451 become very useful to support sensor design and networking. Nevertheless, the importance of sensor networks has been realized by many vendors, and they are promoting proprietary solutions to connect smart sensors into TCP/IP-based networks. Recent developments in the interfacing of sensors are taking place on at least two levels: low-level interfaces and high-level interfaces.

Low-level interfaces focus on the development of sensor clusters consisting of base stations and sensor nodes. It requires Web-enabled smart sensors that that can be coupled directly with the network.

High-level interfaces focus on various wireless network topologies such as Bluetooth, ad hoc networks, Zigbee (i.e., IEEE 802.15.4), Pico radio, and embedded operating systems. More information on these topologies is provided in Section 4.3 and Section 4.4. High-level interfaces minimize the impact of human operators on a wireless sensor networks. There is considerable academic and commercial interest in both types of interfaces.

4.1.3 Effect of IEEE 1451 Standards on Sensor Networks

Manufacturing of wireless sensors for industrial operations requires communication and collaboration between designers and manufacturers to comply with international standards, protocols, and FCC rules. Any wireless sensor standard has to comply with the requirements of the physical layer (e.g., frequency hopping spread spectrum [FHSS], direct sequence spread spectrum [DSSS], and narrowband) and the communication protocols of the upper layers. The standards define universal wireless sensor nodes and their structures. Many organizations such as the IEEE 1451.X Working Group and Bluetooth SIG are making considerable progress toward unification and interoperability.

A common problem in connecting sensors to any wireless network comes from the uniqueness in the design of each sensor and differences in their functionality. The IEEE 1451 family of standards addresses these problems. IEEE 1451 defines the functional boundaries and interfaces necessary to enable smart transducers to be easily connected to a variety of networks.

Transducer manufacturers have long sought ways to build low-cost wireless networks for smart sensors and transducers. Many versions of sensor control networks or fieldbus implementations are currently available, each with its own advantages and disadvantages. Interfacing smart transducers to a variety of control networks and supporting different protocols have proven to be cost prohibitive for users as well as manufacturers. Therefore the transducer interface standard, IEEE 1451, has been useful and very successful in addressing many issues on interfacing and interconnecting sensors. As explained in Chapter 3, the IEEE and the National Institute of Standards and Technology (NIST) have led this activity, which resulted in a family of standards that address various aspects of smart sensor interfacing to networks. One of the major objectives of the IEEE 1451 standard is to make it easier for transducer manufacturers to develop smart devices that are easily interfaceable with the sensor networks, systems, and instruments by simply incorporating existing sensor and networking technologies.

The IEEE 1451 standards define the functional boundaries and interfaces necessary to enable smart transducers to be easily integrated in systems. The standards define protocols and functions that give the transducers interchangeability, self-identification, and network dependence. A key element of IEEE 1451 is the transducer electronic data sheet (TEDS), which allows sensors to self-identify and be interchangeable in networked systems. Using TEDS information, a host computer can recognize a temperature sensor, pressure sensor, or any other sensor type, as well as the measurement range and scaling information.

A smart sensor connected to a node consists of sensing elements, signal conditioning circuitry, and a TEDS. The IEEE 1451 standard clearly defines interface fields within TEDS. Some of the fields that are particularly important for networks are the revision number (1 byte), transducer type (1 byte), details of the manufacturer (17 bytes), sensor model (17 byte), sensor serial number (17 bytes), user alias name (17 bytes), sensor input/output (SIO) identification (2 bytes), calibration type (1 byte), calibration date (7 byte), and next calibration date (7 bytes). These features are either built directly into the sensor or are included in an external microprocessor-based system or SIO devices known as SIO cables. SIO cables are available for sensors with voltage, current, or resistive bridge outputs, as well as resistance temperature detectors (RTDs) and thermocouples. Details of wireless sensor networks can be found in Section 4.3 through Section 4.5.

4.2 Wireless Instrument Architecture and Network Design

Wireless communication of instruments is not new. What is new is the application of new technologies (e.g., IEEE 802, Bluetooth, modulation techniques) that has resulted in the wide proliferation of these instruments.

Modern instruments are made wireless by a number of techniques:

- By embedding wireless features that become part of the instrument.
- By add-on devices or modules, such as transceivers.
- By connecting modems, such as wireless RS-232, wireless USB, etc.
- By using bridges, repeaters, gateways, and data loggers.

It is important to note that the integration of RF communication technology is changing modern instruments, their operational methods, and their application areas. Integration of RF communication systems is preparing instruments as well as users for the next generation of instrument and instrumentation technology.

Embedded wireless instruments consist of five main components: sensors and signal conditioners, programmable digital hardware, memory and storage, I/O and communication components, and other components such as displays, keypads, and power supplies. A typical wireless instrument is illustrated in Figure 4.6. All instruments have similar features to those shown in Figure 4.6, but they differ from each other in the way that signals are handled, transmitted, and displayed.

In embedded wireless instruments, spread spectrum communication techniques are extensively used. Spread spectrum RF technology offers an alternative to the use of wires. The technology in RF communication offers reliability, superior jam resistance, structure penetration, signal sensitivity, and good external signal range. Spread spectrum is available for nonmilitary use and was developed to operate at 915 MHz, 2.46 GHz, and 5.8 GHz frequencies. The transceiver chips complying with spread spectrum capabil-

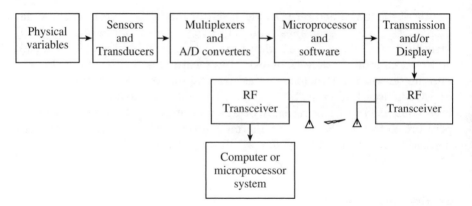

FIGURE 4.6
Components of a wireless instrument.

ities are available for $2 to $10 from companies such as Axonn (New Orleans, LA). These transceivers operate at low voltages (1.8 V to 5 V), hence they have long battery lives (3 to 10 years). Low-cost transceivers are used in distributed systems such as distributed sensor networks, SCADA systems, and so on. An example of a low-cost transceiver is the RCM2200 by Z-World Engineering (Davis, CA). This single-board computer has an Internet interface embedded TCP/IP that allows high-speed network data transfer. The firmware of the RCM2200 can be programmed to support multiple data output formats including hypertext markup language (HTML) for direct Web interface.

Modular wireless instruments use transceivers. Transceivers are complex devices that integrate RF components, including signal generators, amplifiers, attenuators, modulators, demodulators, frequency multipliers, power dividers, power combiners, and signal processing circuits. Two typical examples of transceivers that are used extensively in wireless instruments are the following:

- The SA1638, manufactured by Philips (http://www.philips.com/), is a 48-pin transceiver that combines receive, transmit, and IF circuits. The receive path of the device contains an IF amplifier, a pair of quadrature down-mixers, a pair of baseband filters, and amplifiers. On the transmitter side, a second pair of mixers transposes a quadrature baseband input up to the IF frequency. An external VCO signal is divided down internally and buffered to provide quadrature local oscillator signals for the mixers. A further divider chain, reference divider, and phase detector are added to avoid the need for an external IF synthesizer. The receive or transmit path or the entire circuit may be powered down by logic inputs. On-board voltage regulators with 3.0 V outputs are provided to allow direct connection to a battery supply of 3.3 V to 7.5 V. Current consumption is 18 mA for receive and 22 mA for transmit.

- The CC2420 single-chip transceiver is manufactured by Chipcon (http://www.chipcon.com). This transceiver is designed for low-power, low-voltage RF applications in the 2.4 GHz unlicensed ISM band. It is one of the first commercially available RF transceivers compliant with the IEEE 802.15.4 standard and compliant with 2.4 GHz ZigBee applications. The device is based on a DSSS modem with 2 Mbps and a 250 kbps effective data rate. The packet handling rate is 128 bytes in receive mode and 128 bytes in transmit mode. Media access control (MAC) encryption and authentication is supported by the hardware. The power levels and other characteristics are compliant with EN 300 440, FCC CFR 47 Part 15, and ARIB STD-T66 rules and standards. Current consumption is 19.7 mA in receive mode and 17.4 mA in transmit mode. The output power is programmable and it accepts voltages of 2.1 V to 3.6 V.

Wireless modems are used for wireless communication in conventional instruments that have appropriate I/O functionalities. Since many instruments are equipped with conventional communication ports, such as RS-232, they can be upgraded to wireless communication forms using add-on devices such as wireless RS-232, RS-422 and RS-485, wireless USB, and others. A wide range of equipment for wireless upgrading is available utilizing various technologies such as the IEEE 802.11 family of standards.

An example of the upgrading products that can be integrated with conventional instruments is the DGR-115/115H wireless RS-232 from Airlinx Communications, Inc. (http://www.airlinx.com). It uses spread spectrum data transceivers to provide long-range data communication up to 30 km. By using frequency hopping spread technology, this device is capable of transmitting and receiving uncompressed data at rates of 115.2 kbaud. The DGR-115/115H operates at 1 W output power, thus complying with the FCC Part 15 rules. It operates in either point-to-point or point-to-multipoint modes, selectable through terminal programs. Repeaters may be deployed in either mode to extend the range of the link to 100 km or more.

Another example is the XStream stand-alone radio modem offered by MaxStream (http://www.maxstream.net). This modem provides long-range (30 km), low-power serial communication operating in the 900 MHz or 2.4 GHz ISM bands. It is suitable for connecting to USB or Ethernet ports. The units are dual in-line package (DIP) switchable for RS-232, RS-422, and RS-485 support. It provides interface rates of 1200 bps to 57,600 bps.

Bridges, routers, and gateways are used in sensor and instrument networks. These are explained in Section 4.2.2 and wireless data loggers are discussed in Section 4.2.3.

4.2.1 Essential Components of Wireless Instruments

In addition to the components necessary for conventional wired instruments, wireless instruments require at least two extra components: an RF device and a microprocessor. Since these are critical parts that differentiate wireless instruments from conventional ones, further information will be provided in this section.

Radiofrequency devices are manufactured as ICs using different types of technologies. The available IC technology for RF continues to improve and become more efficient. Today's modern RF systems can contain hundreds or thousands of circuit elements such as transistors, diodes, and resistors.

At present, gallium arsenide (GaAs), silicon bipolar, and bipolar CMOS (BiCMOS) are the major technologies used in the RF market. Despite the disadvantages of high power and high cost, GaAs field effect and hetero-junction devices dominate RF products, especially in power amplifiers and front-end switches. While GaAs processes offer useful features such as higher breakdown voltages and higher cutoff frequencies, semi-insulating substrates, and high-quality inductors and capacitors, silicon technology provides both a higher level of integration and lower overall cost. Silicon technology is suitable for high-volume products with relatively complex circuits, such as frequency synthesizers. Almost all building blocks of typical transceivers are available in silicon bipolar technologies and are offered by many manufacturers.

A competitor of GaAs and silicon bipolar devices is CMOS technology, supported by the momentum of the digital market. RF CMOS devices are produced in large numbers and can operate at high frequencies. CMOS technology is attracting active research interest. However, performance, cost, and time to market are the three critical factors influencing the choice of technology. Amplifiers, mixers, oscillators, frequency synthesizers, and dividers are largely manufactured with both bipolar and CMOS technologies.

Microprocessors are another essential component of wireless instruments. They control the operation of the entire system. Microprocessors that are suitable for wireless instruments are produced by many companies: Motorola, National Semiconductor, Intel, Texas Instruments, Atmel, Hitachi, Cypress, Philips, and many others. Depending on the complexity of the instrument, supporting microprocessors can be selected from three major types: dedicated or embedded controllers, bit slice processors, and general purpose central processing units (CPUs).

Dedicated or embedded controllers are used in relatively simple instruments. Typical examples of embedded controllers are the MSP 430 family of microcontrollers (Texas Instruments), the eight-bit MCS 51 CHMOS family of microcontrollers (Intel), and the 68HC908 series of microcontrollers (Motorola).

Bit slice processors are used in moderately complex instruments. Some examples are the MC2901LC (Motorola), IDM2901A-1DC (National Semiconductor), TS2901BCP (Thompson), and N2901-1I (Signetics).

General purpose CPUs are used in complex instruments. Some examples of general purpose CPUs are the 80219 peripheral component interconnect (PCI) processor (Intel), the TMS320C54x and TMS320C6x family of digital signal processors (DSPs) (Motorola), the RM52XX family of CPUs (PMC Sierra), and Pentium products (Intel). Technical details of some of the microprocessor and microcontrollers are given below.

The Cypress CY8C25xxx/26xxx microcontrollers are built on programmable system-on-chip technology (PSoC) blocks. These analog and digital PSoC blocks implement general purpose microcontroller unit (MCU) peripherals as user modules. A user module is a specific peripheral function (i.e., timing, counting, pulse width modulation, etc.). It is created by setting the personality of one or several analog and/or digital blocks and then adjusting the parameters for the desired function. Using PSoC, one is able to define the desired peripherals by selecting the corresponding user modules. This is done by using the PSoC designer software tool. The advantage of this is that it enables designers to generate "custom" devices. These PSoC blocks can be connected in parallel or serial form to provide different functions. These functions include

- A/D and D/A converters.
- Multipole filters and various gains.
- Timer/counter.
- Serial receiver/transmitter.
- Cyclic redundancy check (CRC) generator.
- Pseudorandom number generator.
- Serial peripheral interfaces (SPIs).

Assemblers and compilers include the M8C language assembler and the PSoC designer C compiler.

The Motorola MC9S12DP256 microcontroller has a 16-bit CPU that consists of

- 256 kb of flash electrically erasable programmable read-only memory (EEPROM).
- 12 kb of random access memory (RAM).
- 4 kb of EEPROM.
- Two serial communication interfaces (SCIs)
- Three SPIs
- Eight channels of input and output capture enhanced timer.
- 10-bit A/D converters.
- Eight channels of pulse width modulation (PWM).
- Digital byte data link controller (BDLC).

- 29 discrete digital I/O channels (port A, port B, port K, and port E), where 20 of them have interrupt and wakeup capability.
- Inter-IC bus.

Assemblers and compilers are Embedded Workbench and ImageCraft ICC12 V6. This microcontroller is able to operate in two different modes—normal and emulation—as well as a special operating mode.

The AT91 ARM Thumb microcontroller family is a 32-bit system. Its architecture is based on the reduced instruction set computer (RISC) processor. The key features of this microcontroller are

- 8 kb of on-chip static random access memory (SRAM) which is directly connected to a 32-bit data bus and is single-cycle accessible.
- Fully programmable external bus interface (EBI) enables connection of external memories and application-specific peripherals.
- Eight-level priority, individually maskable, vectored interrupt controller to reduce the latency time.
- 32 programmable I/O lines where the user can define the lines as input or output. These I/O lines can be programmed to detect an interrupt on a signal on each line.
- Universal synchronous asynchronous receiver transmitter (USART) which permits the user to select the communication mode (i.e., asynchronous or synchronous at a high baud rate).
- Three channels of 16-bit timer counter (TC) which are programmable and able to capture waveforms.
- A tristate mode used for the debugging process. This enables users to connect an emulator probe to an application board without having to desolder the device from the target board.
- Five peripheral registers: control, mode, data, status, enable/disable/status.

Assemblers and compilers are the Multi 2000 development environment, the ARM SDT and ADS development environment, and the IAR Embedded Workbench.

4.2.2 Wireless Bridges, Routers, Gateways, and Repeaters

If the data from wireless instruments needs to be transmitted over long distances which are out of range for the transmitter and receiver, then wireless bridges, routers, gateways, and repeaters are employed. More information on these devices can be found in Section 3.3.3. Some examples are provided here.

Wireless bridges provide long-range point-to-point or point-to-multipoint links. Some bridges use DSSS communication technology, while others use FHSS and other techniques usually operating at frequencies of 2.4 GHz. Data are transmitted at speeds of up to 11 Mbps. These bridges comply with standards such as IEEE 802.11b for connecting one or more remote sites to a central server or the Internet.

An example of a wireless bridge is the DWL-1800 series of devices manufactured by D-Link (http://www.dlink.com/). These devices are comprised of two products, the DWL-1800B wireless base unit and the DWL-1800R wireless remote bridge. The wireless base unit and remote bridge together provide long-range point-to-point and point-to-multipoint links for outdoor applications. They use DSSS technology operating at 2.4 GHz. Data is transmitted at rates of up to 11 Mbps. Both the base unit and remote bridge are 802.11b-compliant devices that connect one or more remote sites to a central server or Internet connection. Both units are equipped with an integrated antenna for outdoor use. The DWL-1800 series can be used as high-speed connections between two or more remote networks in different buildings. Typically a maximum of 128 remote sites can be serviced by the base unit (DWL-1800B) and the DWL-1800R can service up to 1024 stations or connections.

Wireless routers for indoor and outdoor applications use narrowband or broadband technology such as 802.11g. Broadband wireless routers support more bandwidth, allowing more communicating devices to be networked. There are many examples of commercially available routers, such as the MN-700 (Microsoft), 2804WBR (SMC), WRT54G (Linksys), F5D7230-4 (Belkin), M8930LL/A (Apple), and many others. Most of these routers are produced using 802.11g specifications.

Wireless gateways are used primarily in centralized wireless local area network (WLAN) management and control. They have a wide range of applications, such as managing traffic in voice-based systems, connecting remote locations, and providing local and rural connectivity over Global System for Mobile Communications (GSM) networks. A typical example is the WG-400 and WG-1100 series of gateways offered by Bluesocket (http://www.bluesocket.com/). The WG-400 supports up to 50 simultaneous users, while the WG-1100 supports up to 100 users at a message rate of 30 Mbps encrypted and 100 Mbps unencrypted.

Wireless repeaters are employed to regenerate a network signal in order to extend the range of the existing network infrastructure. A WLAN repeater receives radio signals (802.11 frames) from an access point, end-user device, or another repeater and retransmits the frames. This makes it possible for a repeater located between an access point and a distant user to act as a relay for frames traveling back and forth between the user and the access point.

A typical example of a wireless repeater is the 7625 wireless repeater offered by Ambient Weather (http://www.ambientweather.com/). This device is powered by solar power and transmits data to distances of up to 1 km. The transmitting and receiving range for each repeater is up to 120 m outdoors, with line of site. The typical range through walls in indoor applications is 30 m to 120 m.

4.2.3 Wireless Data Loggers

A wireless data logger is a recording instrument that monitors and reports various measurements such as temperature, relative humidity, light intensity, voltage, pressure, or shock. Most data loggers are self-powered, therefore they have mobility while continuously recording information for specific variables. Since they are stand-alone devices, they are used to verify and control the quality of any desired variable. Data loggers are used extensively for measurements involving remote operations, such as petroleum and mineral exploration, environment monitoring, and weather reporting.

Data loggers are available in different sizes and with various capabilities. A single-unit data logger consists of a transceiver capable of communicating locally or over relatively short distances. Since they operate over short distances, the collected information needs to be stored and downloaded to a nearby base station, as in the case of laboratory applications. A multichannel data logger operates over long distances; some use satellite communication systems. These have the ability to interface with multiple sensor arrays in remote locations.

An example of a data logger is a single-chip device integrated with an RF transceiver, as illustrated in block diagram Figure 4.7. It collects information from an onboard triaxial MEMS accelerometer and transmits the information

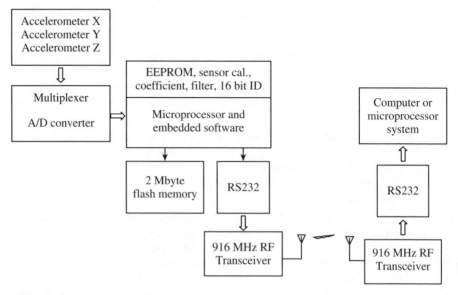

FIGURE 4.7
A typical wireless data logging system.

to a local receiver. In this device, the transmission frequency is 916 MHz, narrowband. The RF communication link operates at 19,200 baud and the device is capable of triggering a sample to be logged (typically from 30 m) or requested data to be transmitted. They have flash memory (typically 2 Mb to 8 Mb. The nodes in the wireless network may be assigned with 16-bit words, hence being able to address thousands of multichannel sensor clusters. This data logger is powered by 3.1 V to 9 V using lithium ion AA-size batteries. The current drain from the battery is about 10 mA at full load. This device contains 10-bit or 12-bit A/D converters. The dimensions of the total package are 25 mm × 40 mm × 7 mm.

An example of a long-range data logger is the WM2M, by Wireless Applications Corporation (http://www.wireapp.com/). This device is capable of using wired and wireless communication with built-in general packet radio service (GPRS) connectivity. It offers short message service (SMS) and circuit-switched data (CSD) connections over 850 MHz, 900 MHz, 1800 MHz, and 1900 MHz GSM networks. It uses a 16-bit microcontroller and allows C-code programming for specific applications.

In other categories, industrial plug-and-play and long-range data loggers are offered by many companies such as Microwave Data Systems (http://www.microwavedata.com/). Typical products provide point-to-point and point-to-multipoint communication of data with frequencies from 200 MHz to 2.4 GHz and speeds up to 8 Mbps. Such devices are used for human health monitoring and other civilian applications, as well as in military and industry uses. Networks of low-cost wireless sensors enable monitoring of buildings and other structures with a large number of sensing nodes.

4.2.4 Power Considerations of Wireless Instruments

A reliable and long-lasting power source is very important in wireless instruments and sensors. A poor power source can significantly limit the operational life and usefulness of the device. During design and construction, a range of power supply options may be available, as there are many different types of power sources. Power sources include chemical batteries, RF energy, vibrational energy, solar power, heat differentials, AC/DC, and so on. Low-power sources such as solar, photovoltaic, heat differential, thermoelectric, vibrational energy, and piezoelectric. These are used in some applications where the devices consume very little energy. For example, RF power works well for RFID products, but their ranges and bandwidths are limited. Using AC/DC line power significantly limits movement in wireless systems. For these reasons, manufacturers largely rely on chemical batteries.

Batteries are electrochemical devices that store chemical energy and turn it into an electrical form. For many wireless instruments, selecting the correct batteries is very important since they are the only source of power. All batteries have the following characteristics:

- They have a limited lifetime and the lifetime is determined by use.
- Batteries provide isolation and offer noise-free operation.
- The output power, capacity, and terminal voltages of batteries are temperature dependent—they increase as the temperature increases.
- Because of internal resistance, the terminal voltage drops when heavy current loads are drawn.
- The output voltage deteriorates with aging, which can be attributed to the deterioration of the chemicals and electrodes inside the battery.
- Excessive current drawn from the battery can cause heating. When there is a large current drain, some batteries, such as lithium and nickel-cadmium types, can be permanently damaged.
- The magnitude of the internal resistances differs. Lithium and nickel-cadmium types have low internal resistances, while zinc-carbon types have high resistances.

Batteries are classified as either primary (nonrechargeable) or secondary (rechargeable) batteries. In both categories, there are many different types based on the materials used and the chemical processes involved. Lithium batteries are known to exhibit good performance compared to other types. In 3.3 V output options, lithium batteries have a high peak output, low self-discharge, low weight, high energy density, good cost:capacity ratio, and small memory effect. For rechargeable primary batteries, lithium ion and lithium thionyl chloride types exhibit superior characteristics.

Rechargeable batteries are used in applications that require frequent battery changes, as in the case of a laptop computer. In the rechargeable range, lithium ion and lithium polymer exhibit good energy density, good peak current, low weight, high voltage, and little memory effect. Lithium polymer is more rugged and environmentally robust, but tends to be more expensive. Lithium polymer batteries can be manufactured in various sizes and shapes. Nickel cadmium and nickel hydride batteries exhibit poor characteristics because of their high self-discharge features, heavier weight, lower energy density, lower voltage, and high memory effect.

In wireless instruments, both primary and secondary battery power have distinct advantages and equally significant disadvantages. Capacity, energy density, discharge service life, shelf life, and discharge rates must be considered. The capacity of a battery is the amount of available energy it provides, which is usually expressed in ampere-hours. Capacity is an important factor in the design of battery packs and recharging systems. The shelf life is the length of time a battery can be stored without loosing so much of its charge that it can no longer function normally, which is typically 50% of full capacity. Since all batteries have internal leakage currents, their voltages slowly decrease over time. This is particularly severe in the case of rechargeable batteries. The energy density is the amount of energy in the battery, measured in watt-hours per unit volume or weight. The energy density depends on the size and weight of the battery. The continuous and momentary discharge rates indicate how much current a battery can supply under various conditions. These discharge rates may be limited by cell polarization. Cell polarization causes the accumulation of chemical products such as gas bubbles. This interferes with the reaction in the electrodes and appears as an increase in the internal resistance. In some cases, the heat produced when a battery supplies excessive current may limit the momentary discharge rate.

Batteries come in many different sizes and the size is directly related to the power consumption and operational requirements of a device. To make effective use of the limited power available, sensor electronic design and operational methodology must carefully be formulated with battery drain as the primary concern. The simple approach of adding a wireless connection to an existing sensor design is not a feasible solution from a battery standpoint, as radios consume large amounts of power to propagate electromagnetic energy.

Power management is important in wireless sensor and instrument applications. The limited power available in batteries requires careful planning for their effective use. To achieve power efficiency, careful circuit design and aggressive battery management policies should be followed. Efficiency can be obtained by selecting appropriate components and imposing methods such as standby modes and duty cycle power management schemes. Time management of operations can produce operational lives tens or even hundreds of times longer than those without. Some of the concerns in power management include

- Average and peak current consumption during both routine and unusual operations.
- Voltage levels that are necessary to drive various circuits.
- Minimum allowable voltage for the device to operate at full capacity.
- How frequently the batteries need to be recharged or replaced.
- Recharge time of the batteries; this could be critical in some applications where continuity in measurements is essential.
- Use of interchangeable battery packs.
- Shelf life of the instrument, which is the expected time the instrument may wait unused but is still able to perform fully within the specifications.
- External temperature ranges of normal expected operations and the response of the instrument to extreme temperature conditions.
- Weight, size, and cost restrictions.
- Integration of a battery charger within the instrument and its cost implications.

- Prediction of unusual circumstances that may affect the performance of the instrument, such as strong light or strong electric, magnetic, or electromagnetic fields.

In wireless instrument and sensor applications, the life of the battery can be extended by the application of suitable techniques. Extended battery life may be critical in many cases, particularly when expenses associated with either replacing or recharging batteries is considered. For example, say a wireless accelerometer is designed to operate on an A-size lithium battery with a capacity of 1000 mAh, with a standby current flow of 3 µA and an operational current of 16 mA. These requirements include the power consumption of the sensor, signal conditioner, microcontroller, memory, and transceiver. For such requirements, the standby life of the battery would be 38 years, but this is not realistic since 38 years of operation is much longer than the shelf life of the battery. Chemical batteries are normally designed to last about three years if they are heavily used.

Let's look at the operational requirements of this particular system. Say the sensor performs 9000 operations and each operation lasts for 20 s. If the wireless sensor monitors eight times a day, this leads to a projected battery life of 3.08 years. If it monitors four times a day, the projected battery life would be six years.

When estimating the battery life of a particular operation in wireless instruments, one must take network traffic and the reliability of the physical layer into account. Additional transmissions caused by errors or collisions will reduce battery life considerably. In contrast, if the sensor operates continuously, it would operate only 50 h (2.08 days). For maximum confidence, wireless sensor systems should be designed to perform periodic self-testing for power fluctuations in the system as well as continuous checking of the battery power level.

4.2.5 Other Wireless Instrument Issues

Communication in wireless instruments takes place on two levels. The first is communication at the sensor level taking place among the internal components of the instrument itself. The second is at the instrument level as it is communicating with other external devices such as peers, computers, or controllers that are part of a network. There are two popular methods for external communication of modern wireless instruments: RF (includes microwave) and infrared (IR).

Radio frequency instrument communication is gaining momentum in high-data throughput technologies such as Ethernet, wireless fidelity (Wi-Fi), and Bluetooth. These technologies allow users to leverage existing infrastructures (via base stations and access points), utilize high-bandwidth flexibility in transmission schemes, and provide adequate networking over distances ranging from a few meters to tens of kilometers.

Infrared communication technology also finds applications, particularly in short-distance operations. For example, in some dosimeters, information exchange between an electronic reader and the dosimeter is provided through IR communication operating through an RS-232 port. When the dosimeter is placed on the reader window, an IR emitter forms signals under microprocessor control according to an exchange protocol. By using suitable software, the operator can register the dosimeter, enter the user's personal information, create a database for the registered dosimeter, transmit the dose accumulation history from the memory to a PC and store this history in a database, view the history stored in a database, transmit date and time, transmit thresholds values, and so on.

Major areas of concern with RF communication are reliability, protocol implementation, and application specifications. Other issues are the cost, power requirements, power efficiency, distance of transmission, interface with other devices, and compliance with FCC rules. Inevitably all designs must carefully balance performance, cost, size, and power consumption. Some of the important issues in wireless instruments are discussed below; power issues are discussed in Section 4.2.4.

Communication reliability is clearly an important issue, as the information moving from one place to another must arrive safely. In wireless systems, reliability is achieved by using modern coding techniques, implementing spread spectrum methods, and using encryption/decryption effectively. Wireless instrumentation networks must operate reliably in environments where RF interference exists from motors, lighting, and other noise generating systems. Most RF noise is caused by bursts of narrowband signals, random electromagnetic interference (EMI) from background noise, and deterministic EMI generated by the RF communication system. Even if the reliability may be low due to adverse environmental conditions, there must be a 100% probability that a message will get through within a reasonable time. It is also necessary to ensure that any errors in the message will be detected and corrected. Lost or corrupt data show up as distortion and can produce false information from the instrument. Collisions between packets and interference from other communication systems and other sources must also be addressed.

Protocol implementation in most RF communication in instruments takes the simplistic view and uses few layers of the open system interconnection (OSI) stack. Systems begin with the physical layer, which is the actual radio circuitry. Many possible combinations of physical layers, communication protocols, and applications exist for wireless communication suitable for instruments. On top of the physical layer, the communication protocol is built using a few upper layers of the OSI model, which defines how information is packeted, routed, and encrypted, and how error checking and corrections are handled. The final layer is the application layer, which defines what information must be transmitted through the communication protocol and what to do with received information once it is extracted from other

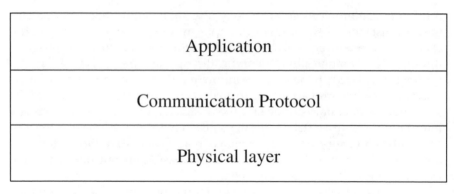

FIGURE 4.8
Layering in digital wireless nodes.

TABLE 4.1

Relative performance of physical layer RF solutions

Physical layer	Cost	Size	Battery drain	Range	Interface rejection
FHSS (Bluetooth)	Medium	Medium	Medium	Low	Medium
DSSS (IEEE 802.11b)	High	Large	High	High	High
Narrowband	Low	Small	Low	Medium	Low
Multiple-frequency narrowband	Low	Small	Low	Medium	Medium

information. Figure 4.8 illustrates a simplified version of the most relevant layers of the OSI reference model.

The potential choices for the physical layer are the narrowband and spread spectrum techniques, both of which yield good results when properly utilized. Narrowband systems can be broken down into two categories: those that use only one frequency and those with multiple frequencies. A well-designed, multifrequency, narrowband radio with spatial diversity performs well in most industrial environments. As discussed in Chapter 2, spread spectrum systems come in two main forms: FHSS and DSSS. The spread spectrum technique improves transmission reliability by distributing (hopping) the transmitted power over a large bandwidth. The hopping rate of FHSS must be slower than the bit rate, but with DSSS, the hopping rate can be faster. In general, spread spectrum systems offer better physical layer performance than their narrowband counterparts, but at the price of increased cost, size, and power drain from the batteries. The relative performances of various physical layer solutions are given in Table 4.1.

One of the potential benefits of spread spectrum transmission is the process gain, which can improve the sensitivity of the receiver, but at the expense of additional battery drain. However, almost all spread spectrum systems used in wireless system networks are FHSS (e.g., Bluetooth), which hop too slowly to take advantage of the process gain. DSSS radio (e.g., IEEE 802.1) also benefits from process gain.

In many applications where multiple nodes are involved, signal reception and transmission can be improved by spatial diversity of the nodes. Spatial diversity refers to appropriate locations of the transmitters and receivers and the use of multiple antennas. Spatial diversity is useful in both narrowband and spread spectrum systems. For example, a simple multiple-antenna design in narrowband receivers can improve spatial resolution considerably by removing the probability of multipath interference. Modern multiple antenna systems provide receiver directional gain through beam forming.

All transmitters in the United States must meet FCC guidelines for licensed or unlicensed operation. Licensed operation may be impractical for most industrial users because of the regulatory hurdles imposed by the authorities, therefore unlicensed operation is preferred whenever possible. For industrial applications, the ISM bands (902–928 MHz and 2.400–2.485 GHz) meet FCC requirements for unlicensed operation. Within the ISM bands, narrowband operation is limited to 1 mW of radiated power, while spread spectrum transmitters are allowed up to 1 W of transmitted power. Higher frequencies provide more signal bandwidth, but lower frequencies offer better performance in multipath environments, where reflections are common.

Application specifications play an important role in wireless instruments. For example, many industrial wireless networks must also allow thousands of sensors to coexist in a single confined space. As discussed in Chapter 2, various techniques are available to ensure dependable coexistence. Some of the common techniques are time division multiple access (TDMA), frequency division multiple access (FDMA), and code division multiple access (CDMA). The circuit cost and complexity of TDMA is lowest, followed by FDMA and CDMA. In the TDMA and FDMA methods, differences in timing or frequency bands are used to separate communication with multiple sensors. In the CDMA method, several communication channels can be open simultaneously, as long as the spread spectrum chirping rates are kept high.

To improve wireless communication, many possible solutions exist. However, any design approach for improvement must carefully balance the performance, cost, size, and power consumption of the products. This balance is particularly important for wireless sensors. For example, CDMA spread spectrum is known to be reliable in data communication, but at a significant increase in cost, size, and power drain. Another solution may be designing less reliable physical layers, but improved communication protocols, to achieve a similar reliability.

In many applications, continuous multiple accesses from a device may not be necessary. For example, for a 256-sensor cell, where each sensor takes readings eight times a day and each communication takes 5 seconds, the total airtime of all the instruments is 2.84 h/day, or only a 12% network loading. Both FDMA and TDMA are viable options for such environments.

With multiple frequencies in narrowband communication with spatial diversity, together with aggressively designed communication protocols, it is possible to achieve enhanced results while still meeting cost, size, and power consumption requirements. In this respect, relaxed timeliness of information flow between nodes can help to improve the system.

4.3   Wireless Sensor and Instrument Network Design

Wireless sensor networks have received considerable attention over the past few years, as they can potentially benefit a variety of scientific, military, and commercial applications. These networks consist of a collection of sensor nodes communicating at RFs and each may incorporate one or more sensors. Nodes may be simple, just taking basic measurements, or they may be complex, containing a large volume of data and information. All nodes must have some kind of signal and data processing capabilities or be under the instructions of local controllers. In many applications, independent sensors make a local decision and then combine these decisions at a fusion center to generate a global decision. In distributed structures, there are a number of ways of combining the decisions of each node, such as providing decision feedback connections to each of the sensor nodes or using sensors that are fully interconnected via a decision feedback mechanism.

In the majority of applications, wireless sensor networks are predominantly data-centric rather than address-centric. That is, queries are directed to a region containing a cluster of sensors rather than to specific sensor addresses. Considering the similarity in the data obtained by sensors in a dense cluster, aggregation of the data is performed locally; that is, a summary or analysis of the local data is prepared by an aggregator node within the cluster, thus reducing the communication bandwidth requirements. Aggregation of data increases the level of accuracy and incorporates data redundancy to compensate for node failures. A network hierarchy and clustering of sensor nodes allows for network scalability, robustness, efficient resource utilization, and lower power consumption.

The design and construction of wireless sensor networks is a subtle process that includes

- Selection of embedded processors for individual nodes and controllers.
- Memory requirements and effective use of memory.
- Communication techniques and protocols.
- Scalable architectures and programming models.
- Routing of information and path finding.
- Robustness at the network layers.
- Operation in real time.

- Safe and uncorrupted transmission and reception of data.
- Data storage, data aggregation, and handling of errors.
- Delays and packet losses.
- Control and network instructions.
- Collaborative signal processing.
- Robustness and security.
- Energy and resource management, including efficiency in power use and power awareness at the application layers and network level.

Real-time wireless sensor networks impose errors, delays, and losses. It is essential that the overall sensor network design consider all the above factors by suitably combining hardware and software. Some of these issues are addressed in the remaining sections of this chapter.

Network topologies for distributed sensors include two basic types: functional hierarchy and decentralized (flat) structure. Depending on the application requirements, a hierarchical or flat structure may be employed. Both flat and hierarchical structures have advantages and disadvantages. The flat structure is more survivable, since it does not have a single point of failure; it also allows multiple routes between nodes. The hierarchical structure provides simpler network management and can help further reduce transmissions. However, the first action for a sensor network to function properly is to determine its topology. This is necessary because many of the traditional routing protocols require topological information for initialization. In many applications (e.g., Bluetooth), although networks start using a flat structure, the network organizes itself in a hierarchical manner, as illustrated in Figure 4.9, in a tree topology.

Communication and robustness depend on the type of structure employed. If communication costs are a primary concern, then the functional hierarchy is preferred because it leads to less network traffic. However, if robustness is the primary issue, then a decentralized structure may be a better choice.

Wireless industrial instrument networks have similar flat or hierarchical structures. In addition, in most industrial environments, instruments are fixed in some locations and can be networked as in the case of cellular telephone systems, by having cells, as illustrated in Figure 4.10. A base station is located at the center of each cell site, which is usually in a hexagonal shape. The cells can be arranged in clusters, for network efficiency.

Cell and supercell arrangements of the base stations partially solve the problem of limited availability of frequency bandwidths allocated for communication. Spacing cells of the same type and reducing transmitter power can prevent interference on other devices using the same frequencies. Special arrangements such as locating the base station at the intersection of three cells and using directional antennae (with sectors of 120°) can reduce the number of cell sites. In addition, not all the cells in a service area carry the

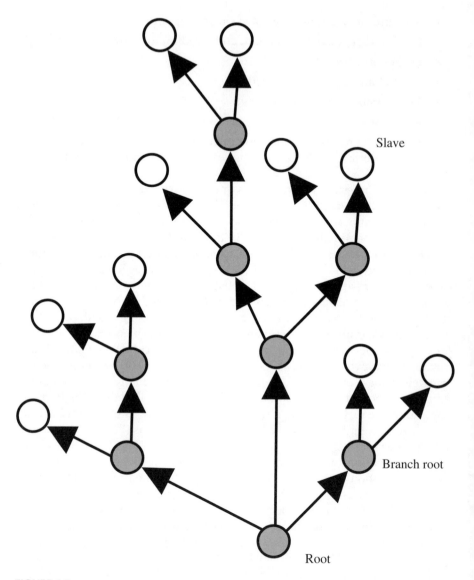

FIGURE 4.9
A sensor network with tree topology.

same traffic. Cells in heavily populated areas are made smaller to allow connections of more instruments.

4.3.1 Mobile Wireless Instrument and Sensor Networks

The term mobile has historically been used to classify any radio terminal that could be moved during operation. More recently the term mobile is

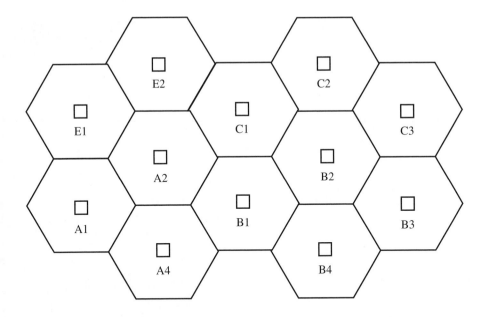

A, B, C,....Cell clusters and 1, 2, 3,....Frequency groups

FIGURE 4.10
Cells and cell cluster arrangements of wireless instruments.

used to describe a radio terminal that is attached to a high-speed mobile platform (e.g., cellular systems), whereas the term portable describes a hand-held radio terminal that can be used by someone while he or she is moving about. A portable instrument is a self-powered device and can be used in different locations without the need for an external power source. A mobile wireless instrument is a portable instrument that has been equipped with wireless communication capabilities. A portable instrument can assume measuring and monitoring tasks while it is moving and can also send and receive information.

Figure 4.11 illustrates a simple wireless communication system that can be applied in wireless sensors and instruments. This system includes a base unit and mobile transmitter and receiver. In this system, a full-duplex scheme is employed such that the transmitter and receiver are on slightly different frequencies in the operation band of interest. Since a duplex system uses different frequencies for transmitting and receiving, it is possible to transmit and receive at the same time. In such systems, a digital security coding scrambler is used to prevent interference from other nearby devices.

Mobile wireless sensor and instruments networks are good examples of flat sensor network structures, often called ad hoc networks, as illustrated in Figure 4.12a. In mobile systems, sensors shift from one location to another in an ad hoc manner. As the sensors move, as in Figure 4.12b, the network topology changes, requiring the network to periodically reconfigure itself.

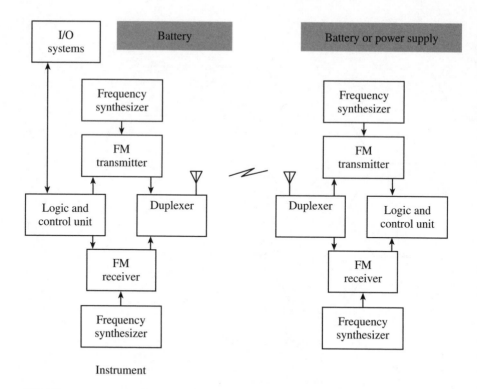

FIGURE 4.11
The structure of a typical wireless instrument.

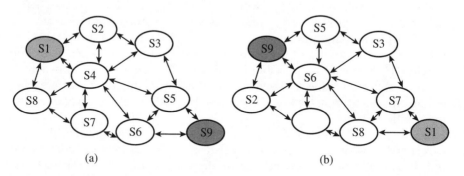

FIGURE 4.12
An ad hoc sensor network.

To be able to reconfigure the topology, routing protocols must be able to handle this changing situation. Most ad hoc routing protocols can be classified into two categories: proactive and reactive. Proactive and table-driven routing protocols maintain routes to all possible destinations by periodically exchanging control messages. Reactive and on-demand protocols, on the other hand, discover routes only when there is a demand for it. The choice of which system to use depends on the network algorithms implemented.

Also, the media access mechanism of the communication process must be adaptable for the new topology. For example, sensor 9 (Figure 4.12), communicating with sensors S5 and S6, needs to establish new links with sensors S2, S6, and S5. In the reconfiguration process, link parameters such as modulation type, amount of channel coding, and transmitter power must adaptable. Moreover, the distributed detection and estimation algorithms must also be flexible.

4.3.2 Energy Issues in Wireless Sensor Networks

Wireless distributed sensor networks may consist of hundreds to thousands of small sensor nodes scattered throughout the area of interest. Each node collects data and shares information about that data with other nodes. The entire network is designed to operate in a collaborative, maintenance-free, and fault-tolerant manner. Often nodes are required to operate in an energy efficient manner, optimizing computation and communication needs using as little energy as possible. In energy critical applications, all levels of communication, from the physical and link layers to the application layer, must be configured for energy efficiency. To achieve this, a total system approach is required for efficient and reliable self-powered sensor networks. Issues such as standby leakage energy and startup overhead must be considered.

A version of a wireless sensor network is achieved by using microsensors. The microsensor node integrates the sensing, processing, and communication subsystems together in a single chip or on a single, small and compact circuit board. In design and construction, several options may be available; for example, the operational nodes can be configured using low-power commercial off-the-shelf (COTS) components. Figure 4.13 illustrates a self-powered, COTS-based sensor node operating on the 2.4 GHz Bluetooth standard.

Energy dissipation in microsensor systems is critical, and energy efficiency can be obtained in several ways, including

* Shutting down inactive components through operating system control.
* Dynamic voltage and frequency scaling of the processor core.

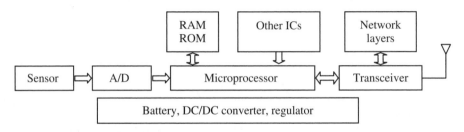

FIGURE 4.13
A commercial off-the-shelf-based sensor node.

- Using carefully developed processing algorithms.
- Employing energy-aware computing.
- Controlling idle mode leakage currents and low-power modulation.
- Designing adjustable and low duty cycle radio power transmission.
- Optimizing the energy in multihop communication schemes.
- Scavenging energy from the environment.

4.3.3 Bluetooth Sensor Networks

Bluetooth is a good example of an ad hoc wireless network. Bluetooth is a low-cost and short-range wireless technology that provides many communication functions. Bluetooth has a range of about 10 m and has a maximum throughput of 1.5 Mbps. Bluetooth's initial aim was to provide short-distance wireless interfaces, mainly to personal area network (PAN) products such as headsets. Today Bluetooth technology has firmly established itself in the marketplace, offering devices fulfilling a diverse range of functions. Devices compliant with Bluetooth technology range from pulse oximeters, to gaming devices, to networking of PC devices such as printers and keyboards.

Bluetooth technology is also gaining wider acceptance in the implementation of wireless communication in industrial instrumentation systems and networks, finding applications in office equipment, home security systems, the automotive industry, PC networks, service tools, medical applications, home appliances, handheld instruments, and many others. For example, Dolphin 9500 and 9550 series mobile computers, by Hand Held Products (http://www.handheld.com/), are used for data collection applications. These devices are integrated with wireless communication systems compliant with Bluetooth, 802.11b, and GSM/GPRS.

Networking Bluetooth devices is relatively easy, as its specifications define how Bluetooth devices can group themselves for the purpose of communication. A Bluetooth ad hoc network can be established by interconnecting devices forming small- or large-scale networks. In small-scale applications, a typical Bluetooth wireless personal area network (WPAN) consists of predefined piconet structures. Each piconet is a cluster of up to eight Bluetooth devices. One of the devices in the cluster is designated as the master, and the others assume the slave position. Two piconets can be connected through a common Bluetooth device. This common device serves as a gateway or bridge. Two or more piconets form a scatternet. Interconnected piconets within the scatternet form the backbone of larger networks such as mobile area networks. The advantage of mobile area networks is that they enable devices to communicate with each other even if they are out of range of each other. This communication can be realized through several hops.

A piconet is an ad hoc and spontaneous clustering of Bluetooth devices. A piconet is formed when two Bluetooth devices enter into communication with each other and determine the master-slave hierarchy, which is formed automatically. One assumes the master position and selects the frequency, frequency hopping sequence, timing of hops, and polling order for the slave, and they communicate within the assigned parameters. If there are more Bluetooth devices in range, the master takes control of the others by assigning frequency, hopping sequence, timing of hops, polling, and so on, for each device.

Any Bluetooth device can function as a master, a slave, or a bridge on a temporary basis as long as the piconet exists. In a piconet cluster, a maximum of seven devices can be active slaves at any given time. If there are more than seven devices, the rest of the slaves cannot be active and they are said to be parked. The maximum number of parked slaves can be 255 per piconet. Each parked slave is assigned a dedicated direct parked slave address. However, indirect addressing of parked slaves by their specific Bluetooth addresses is also permitted, thus allowing any number of parked slaves within range. The number of parked devices therefore can exceed 255 devices per piconet.

Each Bluetooth device has a unique global identification that is used to create hopping patterns. The master device shares its global identification and clock offset with each slave. The slave re-creates the frequency hopping sequence of the piconet it has joined and synchronizes itself with the master's clock. If a device acts as a bridge, it communicates with all piconets connected to it by aligning itself with the clock of each piconet it belongs to. A device assuming a bridge position can communicate with only one piconet at a time, and thus has the potential to be a bottleneck. A bridge may be a slave in all of the piconets or a master in one piconet and a slave in the others. Interconnection of two or more piconets via bridge devices results in the formation of a Bluetooth scatternet.

A Bluetooth device can assume many other tasks, including standby, inquiry, page, connected, hold, park, and sniff. A standby device is powered on but has not yet joined a piconet cluster. In inquiry, the device is seeking other devices it can collaborate with to form a network. In paging, the device is already in a master position and is actively seeking other potential devices, prompting them to join the piconet. A connected device is accepted in the piconet and all the necessary communication links have been established. A slave assumes the transmit position when sending data to the master. At the end of the transmission, it returns to the connected state until it has more data to send or it has been requested by the master to respond. The sniff state is a low-power state in which the slave sleeps for a predetermined time. In the hold state, the slave is not active for a predetermined period of time and it does not communicate until that time lapses. In the park state, the slave has no data to receive or transmit. In the parked state, the master advises the slave to relinquish its active address and the device becomes inactive. The active relinquished address can then be used for other devices that may require parking.

A Bluetooth scatternet is an ad hoc network consisting of two or more piconets connected together by bridge nodes. Piconets coexist in the same area by using different hopping sequences. The construction of ad hoc networks is generally fairly complex. As a result, several models for constructing Bluetooth scatternets have been proposed. The most prominent of these models are BlueTree, Tree Scatternet Formation (TSF), and BlueStar.

The BlueTree protocol is based on the formation of a tree network topology, shown in Figure 4.9. There are many different types of BlueTree protocols. In one type, in the initial formation of a tree, a node is designated as the root of the tree to be formed. Once the root is decided, this node acquires its direct one-hop neighbors as slaves. Each slave then pages its one-hop neighbors that are still unconnected and attempts to acquire them as slaves. This process continues until all the devices in the vicinity are acquired and a tree is built. The tree is then optimized by limiting each master to a maximum of seven slaves.

In another type of BlueTree protocol, several initial roots are formed and each root acts independently in the way explained above. Once several trees are formed, the trees generated by each root merge to form a single tree.

The formation and maintenance of BlueTree requires considerable overhead. For information to flow from the bottom of the tree to the top, or vice versa, a considerable amount of routing and timing may be required. There are several variations of BlueTree, including one that is based on determining positions using global positioning system (GPS) information.

The TSF protocol assigns master and slave roles as it connects each node to its tree structure. TSF consists of one or more rooted trees constantly attempting to merge into a topology with fewer trees. This is a decentralized and self-healing process, in that each tree continuously searches for other trees and nodes. In TSF, a node can enter or leave a tree at any time, but any node that is a member of two or more trees can act as an agent to form larger trees. Communication between nodes is achieved by two proposed states: FORM and COMM. In the FORM state, a device is attempting to detect nodes in other trees with the purpose of forming larger trees. In the COMM state, a device is actively transmitting data.

BlueStar consists of a three-phase protocol for multihop scatternet formation. The first phase is device discovery by the inquiry, page, and connect process, as explained above. The second phase is piconet formation, in which a master is assigned by the use of weights. The third phase is scatternet formation, shown in Figure 4.14. In this stage, the master selects which nodes are to become bridges to connect with other piconets. The selection of a bridge depends on the weight of that node in relation to other masters, therefore a candidate must have communicated with other masters beforehand. BlueStar offers distributed solutions and produces a mesh-like network. A scatternet has multiple routes between pairs of nodes. For example, in Figure 4.14, node 4 can connect node 24 via several piconets (e.g., nodes 1, 17, and 21, or nodes 1 and 13).

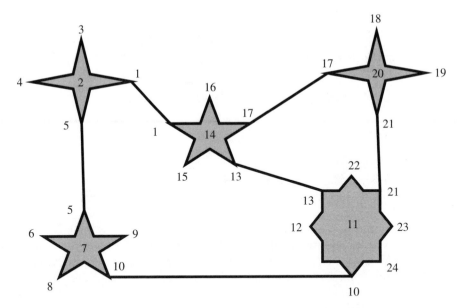

FIGURE 4.14
BlueStar interconnection to form a scatternet.

4.3.4 Applications of Bluetooth Networks

IEEE 1451-compliant smart transducer nodes are integrated with Bluetooth networks, thus signifying an important step forward in wireless sensor and instrument applications. Successful integration requires detailed study of the network communication models that are specified for smart transducer communication as well as the Bluetooth protocol stack. Recently the interface between IEEE 1451 and Bluetooth communication hardware has attracted considerable attention and there is much literature available on this topic.

The IEEE 1451 family of standards can be divided into two basic parts: defining a set of hardware interfaces for connecting transducers to a microprocessor or an instrumentation system, and defining a set of software interfaces for connecting transducers to different networks while utilizing existing network technologies. To connect transducers to different networks, the NCAP of IEEE 1451.5 has significant processing flexibility and is capable of connecting a diverse range of access points to the network. It is implemented in several different ways using wireless technology:

- A minimal NCAP is implemented on a radio module to connect a single smart transducer interface module (STIM) by a wired connection, as in Figure 4.15a.

- A minimal NCAP is implemented on a radio module to connect a small number of STIMs.

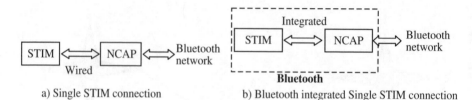

FIGURE 4.15
Integration of STIMs with Bluetooth.

- An NCAP with significant processing capability connects many wireless STIMs to the network in an integrated form, as in Figure 4.15b. This technique allows the NCAP to act as an access point.

An example of an IEEE 1451 Bluetooth device is the CN1000, by Eagle Technology (http://www.eagle.co.za/), a four-channel Bluetooth wireless node. It collects data from multiple sensors and transmits the information via Bluetooth wireless communication to a network hub, computer, or hand-held device. Each channel is ready to receive a fully conditioned analog sensor signal from a smart I/O (SIO). The CN1000 translates the signal received from the sensor into engineering units using the IEEE 1451 TEDS supplied by the SIO. The sensor signal is then digitized by the 16-bit A/D converter for transmission along with the TEDS for each sensor. This allows each channel to identify itself to the host system. The node operates either from an external power supply or an attached battery.

Integrated circuits that support Bluetooth products are appearing in the marketplace. Some of these products are offered in the form of single chips, while others are produced as fairly complex and sophisticated devices. A typical example of chip-level products is the Bluetooth compatible chipset from Texas Instruments. The BSN6030 offers a read-only memory (ROM)-based Bluetooth baseband controller that includes a fully integrated Bluetooth software stack. The TRF 6001 Bluetooth RF transceiver chipset provides up to 1 Mbps for fast data transmission.

Another example of a chip-level device is the STLC2150 transceiver by STMicroelectronics (http://www.st.com/). The STLC2150 provides an on-chip VCO that offers self-calibration capabilities. This chip implements a low-IF receiver for Bluetooth modulated input signals that does not require any external IF filtering. The transmit section features a fully integrated GFSK modulator, followed by a direct up-conversion stage. The STLC2150 can be paired with a baseband processor, such as the STLC2410 or STLC2415. For example, the STLC2410 is a baseband processor that can handle up to seven slave nodes and supports a voice channel through a pulse code modulation (PCM) interface. It also delivers encryption and ciphering capabilities. It was developed around an ARM processor and sports 64 kb of on-chip RAM. Bluetooth offers 128-bit encryption for data security.

The single-chip solution of Bluetooth enables it to be used in unusual applications, such as headsets and luggage security tags. It can also be used in phones, cameras, portable games, and for linking devices to PCs. It is in wide use by companies such as IBM, Sony, Compaq, 3Com, NEC, Fujitsu, and TDK. A typical example of the use of Bluetooth on the device level is in Sony products. In Europe and Japan, Sony has introduced an ultracompact fixed-lens camera (DSC-FX77) with built-in Bluetooth communication.

Many chip-level products target components and systems within Bluetooth specifications. Products range from point-to-multipoint communication components consisting of baseband controllers with flash memory, reference crystals, and RF CMOS transceivers. In some cases, two chips are used to ensure a good level of performance and reliability in RF-intensive and noisy environments. The architecture is based on independent silicon optimization, with digital circuits in standard CMOS and analog parts in BiCMOS or RF CMOS, enabling cost and size reductions.

Another example at the device level is the BlueSerial Bluetooth serial port adapter, by G&W Instruments (http://www.gw-instruments.de/). The Bluetooth PCI adapter is connected to an RS-232 serial port of a computer or other device. This makes the device operate as a Bluetooth node, thus enabling wireless communication with other Bluetooth devices such as other computers, personal digital assistants (PDAs), cell phones, and other instruments. The Bluetooth PCI adapter comes with software support for Windows 98/ME/NT/2K/XP and Linux.

Bluetooth has not escaped competition. ExtremeTech reports that Sony and Philips have agreed to jointly work on "near-field communication," a potential competitor to Bluetooth in the short-range PAN market. Also the IEEE 802 standards are strong competitors of Bluetooth, particularly for longer range communication applications. However, IEEE 802 has incorporated the Bluetooth protocol under its WPAN standards.

4.3.5 IEEE 802 Sensor and Instrument Networks

To address the needs of wireless networking, the IEEE Standards Association defined the IEEE 802 family of standards for local and metropolitan area networks. The IEEE 802.11 family of standards is a wireless version of the IEEE 802.3 standard for wired Ethernet LANs and they are used extensively in sensor and instrument networks. As explained in Section 3.5.1 and Section 3.6.5, these standards are configured to address the current needs of all types of wireless networks. Some network standards such as 802.15.4 (supported by the ZigBee industry consortium) for large sensor networks and 802.16 worldwide interoperability for microwave access (WiMax) broadband access systems are emerging as key technologies. IEEE 802.20 is a proposed system that will support broadband mobile data solutions, to complement third-generation cellular systems. Today these standards have significant momentum in the market.

The IEEE 802.11 standard is designed to optimize throughput, roaming, and distance in communication and connectivity. There are three main sub-standards: 802.11b, 802.11a, and 802.11g. 802.11b offers a bandwidth of up to 11 Mbps using 2.4 GHz, 802.11a and 802.11g offer up to 54 Mbps. Important characteristics of these standards in wireless sensor and instrument networks is that 802.11a can only work at a distance of about 10 m, while 802.11b can reach to 100 m.

IEEE 802.15.4 defines a low-level radio interface for a network that is capable of transporting data through areas of high electrical noise and metallic interference at nominal distances up to 100 m. In addition to its noise rejection abilities, the standard stipulates the use of mesh networking to overcome direct line-of-sight obstructions and to provide alternative path routing in case of temporary network outages. Mesh networking also provides a convenient way to expand coverage distances for ZigBee networks, because the distance limit only applies to the most distant unit. The ZigBee protocol provides the necessary mechanism for removing redundant messages received from alternate paths in the mesh.

The IEEE 802.15.4 standard already occupies a significant portion of the commercial wireless network market. Some of its applications are in building management, factory automation, home automation, wireless sensor networks in factories, the automotive industry, automatic meter reading, and toys. There are few commercial implementations of ZigBee. Chipcom announced its silicon devices supporting ZigBee, and Motorola announced its 802.15.4 solution. Both operate in the ISM 2.4 GHz band. Motorola uses its M68HC08 microcontroller family with its RF packet radio chip. The 802.15.4 and ZigBee protocol stacks are in software. Several other companies, including Intel, Motorola, Atmel, and Phillips, are in the process of developing similar products.

The IEEE 802.16 WiMax standard is designed for wireless broadband access and can reach 50 km or more at up to 70 Mbit/s. The Wi-Fi family (802.11a, b, and g) is designed for wireless short-range Ethernet applications. And Bluetooth, a subset of the 802.15 WPAN standard, has a range up to 10 m, although this can be extended at the expense of increased power consumption, and a data exchange rate of about 1 Mbit/s.

The IEEE 802.15.4 standards define star, tree, and mesh network topologies. This standard mainly addresses two layers of the OSI reference model: communication radio (physical layer) and protocol (data link layer) for both star (point-to-point) and peer-to-peer topologies. ZigBee is used for industrial automation, home control, and building automation.

There are many applications of the IEEE 802 standards at the chip level; for example, the CC2420 RF transceiver, by Chipcon (www.chipcon.com/). This transceiver is backed up by a low-power eight-bit flash microcontroller from NEC (www.nec.com/). This module is compatible with IEEE 802.15.4 and ZigBee wireless network standards for use in a wide range of industrial, commercial, and residential applications. The CC2420 is a 2.4 GHz RF transceiver chip that enables wireless communication via a physical (PHY) layer and media access (MAC) layer that comply with existing 802.15.4 specifications. The NEC microcontroller supports flash programming and debugging via USB connection to a PC.

Another chip-level application is the JS24Z121, an IEEE 802.15.4/ZigBee single-chip device by Jennic (http://www.jennic.com/). This chip provides a single-chip solution for applications requiring microcontroller functionality with a built-in wireless link. It is compliant with the IEEE 802.15.4 standard and provides spread spectrum communication with advanced encryption standard (AES) data flow. It allows applications to be developed using either ZigBee or proprietary networking layers. Wireless functionality and a range of analog (A/D converter, D/A converter, comparator, temperature sensor) and digital (SPI port, universal asynchronous receiver-transmitter [UART], timer, general purpose I/O) peripherals are integrated in the chip. The device is equipped with low system power consumption, particularly in sleep mode. IEEE 802.15.4 software is supplied with the device and the ZigBee protocol stack is available.

At the device level, Tmote Sky, by Moteiv (http://www.moteiv.com/), is an example of an IEEE 802.15.4 compliant device for wireless mesh networking. It is a low-power, high data rate device for networking sensors using IEEE 802.15.4 standards. Tmote offers a number of integrated peripherals, including a 12-bit A/D and D/A converter, timer, I²C, SPI, and UART bus protocols, and a direct memory access (DMA) controller. It provides a USB protocol for programming, debugging, and data collection.

Another example at the device level is the NC220W/NC200 wireless/wired network camera server, by Hawking Technology (http://www.hawkingtech.com/). This camera server allows remote access from a Web browser or Windows utility for live image viewing. This device transmits images at 11 Mbps on an IEEE 802.11b WLAN.

It is expected that soon there will be millions of devices compliant with the IEEE 802 standards.

4.4 Wireless Integrated Network Sensors

Wireless integrated network sensors (WINS) provide distributed network and Internet access to sensors, controls, and processors that are embedded in equipment, facilities, and the environment. WINS development was initiated in 1993 at the University of California, Los Angeles; the first generation of field-ready WINS devices and software was demonstrated three years later. Later the DARPA-sponsored low-power wireless integrated microsensors (LWIMs) project demonstrated the feasibility of multihop, self-assembled wireless networks. This first network also demonstrated the feasibility of algorithms for operating wireless sensor nodes and networks at micropower levels.

Today, the basic network structure and supporting electronic components for WINS are not much different than for wireless sensor networks, as explained in Section 4.3. Therefore details of the components to configure WINS and network topologies will not be repeated here. Instead, the underlying characteristics and surrounding issues for WINS will be highlighted.

Wireless integrated network sensors represent a relatively older technology when compared to wireless sensor networks and wireless mobile sensor systems. WINS were originally intended to facilitate monitoring and control in large-scale applications such as the transportation industry, manufacturing, health care, environmental monitoring, security systems, border control, and traffic control in metropolitan areas. WINS aim is to combine sensor technology, signal processing, decision making, computation, and wireless networking in a compact low-power system. The main feature of a WINS is its RF communication capability, which generally, but not always, provides bidirectional network access for low bit rate, short-range communication of sensors and instruments among themselves and with other host devices.

Most WINS aim to combine sensor technology with a low-power sensor interface, signal processing, data conversion, computing, control functions, and RF communication circuits. The need for low cost and low power consumption can present significant engineering challenges in the design and implementation of these systems, particularly if conventional digital CMOS technology is selected as the building block of the system. In some applications, the sensor component may be required to operate continuously for event recognition, while the network interface of the component may be operating at low duty cycles. This type of operational boundary imposes further challenges in design and applications.

In some systems based on wireless integrated sensor architecture, sensors are required to detect events continuously. All the components, sensors, data converters, buffers, and other elements operate at a micropower level. Upon detection of an event, the microcontroller issues a command for signal processing. Protocols for node operation then determine whether a remote user or neighboring WINS node should be alerted. The WINS node then supplies an attribute of the identified event, for example, the address of the event from a lookup table stored in all network nodes.

Wireless integrated network sensors must be able to support large numbers of sensors in a local area by means of short-range and low average bit rate communication devices. Hence the network design must consider servicing dense sensor distributions with an emphasis on recovering information generated by the application. The architecture therefore exploits the small separation between nodes to provide multihop communication.

Multihop communication (Figure 4.16) provides large power and scalability advantages for WINS. In this particular case, node data are transferred over the asymmetric wireless links to an end user or to a conventional wired or wireless IP network service through a network bridge. However, RF communication path loss has been a primary limitation for wireless networking, with received power (P_R) decaying with transmission range (R) as

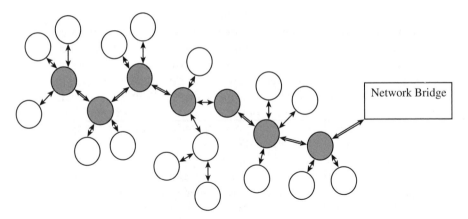

FIGURE 4.16
Multihop communication of WINS.

$$P_R \propto R^{-a}. \tag{4.1}$$

In dense WINS, multihop architectures permit communication links between the nodes, providing advanced capabilities for narrowband devices and allowing a power reduction in the implementation of dense node distributions.

Recent communication and networking protocols for embedded radio is a fast-developing area attracting considerable research effort as applied in WINS. Embedded radio networks include spread spectrum signaling, channel coding, and multiple access techniques such as FDMA, TDMA, and CDMA network protocols. The operating bands for embedded radio are generally selected in the unlicensed bands at 902–928 MHz and 2.4 GHz. These bands provide a compromise between the power cost associated with high-frequency operation and the penalty in antenna gain reduction with decaying frequency for compact antennas. There are some prototypes implemented on WINS with self-assembling, multihop FDMA and TDMA network protocols.

Smart sensors and convenience in signal processing create new opportunities for wider applications of WINS. These opportunities depend on the development of scalable and low-cost sensor network architectures as well as cost-effective RF communication technologies. This requires that the sensor information be conveyed to the user with low-power transceivers. For example, in environmental monitoring applications, continuous sensor signal processing must be provided to enable constant monitoring of events. This necessitates local processing of the distributed measurement data using low-cost and scalable technology as much as possible. Distributed signal processing and decision making allows events to be identified by remote sensors. The information obtained locally is conveyed in short message packets to a host station to take global decisions.

The primary limitations on the WINS nodes are the cost and the power requirements, as most sensors are powered by batteries. In this respect, low-power sensor interfaces and signal processing architectures and circuits enable longer lasting operational systems. However, wireless communication energy requirements present additional demands. Conventional wireless networks are supported by complex protocols that are developed for voice and data transmission for handheld and mobile terminals. These networks are also developed to support communication over long distances with a link bit rate of more than 100 kbps.

Wireless integrated network sensors may operate longer supported by CMOS micropower components such as low-power A/D converters. Taking an A/D converter as an example, the power requirements can be reduced by selecting a low-power A/D converter drawing low current (e.g., 30 μA or less) at full operation. In this case, sample rates for a typical microsensor application may be limited (say to 1 kHz). Also, it is important to note that in some applications the signal frequency may be low—close to DC—as in the case of thermopile infrared sensors, therefore the A/D converter must be highly stable and at the same time must be available at low cost. For this purpose, delta-sigma converters may be suitable. A first-order CMOS delta-sigma converter can have nine-bit resolution for a 100 Hz signal with a power consumption of about 30 μA on a single 3 V rail.

Noise causes problems in many applications. Being low-power systems, WINS require the detection of signal sources in the presence of environmental noise. All source signals decay in amplitude with radial distance from the source. To maximize the detection range, sensor sensitivity must be optimized. In addition, due to the fundamental limits of background noise, a maximum detection range exists for any sensor. This is critical to obtain the greatest sensitivity and to develop compact sensors that may be widely distributed. MEMS technology provides an ideal path for implementation of these highly distributed systems.

4.4.1   Power Requirements in WINS

Wireless integrated network sensors require low-cost, low-power, short-range, and low-bit RF communication to support the entire system. In today's technology, many RF modems are made as embedded radios added to compact microdevices without significantly impacting cost, form, or power. Some examples of these types of wireless sensors were discussed in Section 4.3.

Power for sensor nodes is generally supplied by compact battery cells. Total average system needs must be low (say 30 μA) to provide a long operating life from a typical compact lithium coin-type cell. These cells provide a peak current of no more than 1 mA. Both average and peak current requirements present unique challenges for circuit design. Low-power, reliable, and efficient network operation may be obtained with intelligent sensor nodes that include sensor signal processing, control, and a wireless network interface. A distributed network sensor device continuously monitors multiple sensor systems, processes the sensor signals, and adapts to changing environments and user requirements while making decisions on measured signals. For low-power operation, network protocols must minimize the operation duty cycle of high-power RF communication systems.

Conventional RF system design for WINS is based on a combination of integrated chips and board-level components that employ interfaces between components that drive 50Ω resistive loads, which is required for matching the off-chip transmission lines and components. However, by integrating active and passive components in a single package, impedance may be raised, reducing power dissipation. Impedance within component systems and between component systems is controlled by the introduction of high-Q inductors at each node that balance the parasitic capacitance that would otherwise induce power dissipation. The introduction of high-Q inductors enables narrowband, high-output impedance, weak inversion MOS circuits to be translated from low frequency to an equally narrowband at high frequency.

In RF design, conventional CMOS VCO circuits provide degraded performance at the desired micropower level. However, for VCO systems operating with an LC resonator, having a complete circuit quality factor Q, the advantage in phase noise power is Q^2. This phase noise advantage recovers the performance loss associated with power reduction. But in addition, high-Q resonators, providing voltage gain in the oscillator feedback loop, also allow for a reduction in transistor transconductance. This also results in a reduction in the power required to sustain oscillation.

The introduction of high-Q resonators in embedded radio systems provides the advantage of power reduction. However, this narrowband operation also creates a need for precision in passive component values and the need for tuning. Tunable elements can be based on varactor diodes implemented in the CMOS process.

The inductors required for embedded radio may be implemented in either on-chip elements or passive off-chip components. In on-chip LC circuits, due to substrate and conductor losses, these inductors are limited to Q values of three to five at 1 GHz. These circuit implementations are well suited for broadband, high data rate wireless systems. However, the embedded radio system generally requires narrowband operation and must exploit high-Q components. There are various ways of designing on-chip high-Q inductor systems. Low-temperature cofired ceramic (LTCC) technology, for example, provides flexible device geometry and integration of a flip-chip die attached on a substrate with embedded capacitor and inductor passives. In this design, both primary and secondary coupling coils can be included where the coupling coil provides a 50Ω measurement port.

The LTCC substrate provides low-loss passive components, as well as package support for integrated sensing, signal processing, and microcontroller devices. At 1 GHz, the scale of integrated inductors implemented in

CMOS technology dominates the area of the physical circuit die. However, an inductor implemented in the LTCC substrate requires no die area.

Embedded radio development for WINS is realized by CMOS circuit technology, permitting low-cost fabrication along with the additional wireless components. Many challenges accompany the development of RF systems in the CMOS technology. Of particular importance to embedded radio are the problems associated with low transistor transconductance and the limitations of passive RF components. In addition, WINS embedded radio design must address the peak current limitation of typical battery sources, typically 1 mA. This requires implementation of RF circuits that require one or two orders of magnitude lower peak power than conventional systems. Due to the short range and low-bit characteristics, the requirements for the input noise figure may be relaxed. In addition, channel spacing for the embedded radio system may be increased relative to conventional RF modems, relaxing further the requirements on selectivity. Constraints on operating requirements must consider resistance to interference by conventional spread spectrum radios occupying the same unlicensed bands.

Transceiver power dissipation in conventional RF modems is dominated by transmitter power. In the limit of low transmitter power, less than 1 mW to 3 mW, for WINS, the receiver system power dissipation equals or exceeds that of the transmitter. This is a direct result of the increased complexity of the receiver, the requirement for power dissipation in the first stage amplifier, and the power dissipated by the VCO. Therefore it is critical to develop methods for the design of micropower CMOS active elements. These circuits must operate in the MOS subthreshold region at low transconductance.

4.4.2 Application of WINS

Although intended for large-scale applications, compact geometry and low-cost allows wireless integrated sensors to be embedded and distributed at a small fraction of the cost of conventional wired sensor and instrumentation systems in indoor applications. Wireless integrated sensors also find applications in

- Monitoring of land, water, and air resources.
- Monitoring and control of transportation systems and borders for efficiency, safety, and security at the national level.
- Battlefield situational awareness that provides personnel movement and health monitoring to enhance their security and efficiency at the local level and wide-area scale.
- Control of traffic in metropolitan areas and for security, emergency, and disaster recovery services.
- Creating a manufacturing information service for cost and quality control in organizations and enterprises.

- Sensing, monitoring, and controlling functions in biomedicine applications to connect patients to clinics, ambulatory outpatient services, and medical professionals.
- Maintaining condition-based devices and equipment in power plants, factories, appliances, vehicles, and energy systems to enhance reliability, improve the quality of services, and reduce energy use.

An important requirement for the WINS node appears for conventional sensors and micropower sensor interfaces. For many applications (e.g. military), the WINS system needs to operate at low power, with sampling at low frequency and with limited environmental background sensitivity. The micropower interface circuits sample at DC or low frequency where "$1/f$" noise in those CMOS interfaces is large. The micropower signal processing system can be implemented at low power and with limited word length. The WINS network supports multihop communication with a wireless bridge connection to conventional wired network services.

Recently, complete prototype WINS networks have been demonstrated in defense, factory automation, and condition-based maintenance applications. These results reveal a wide range of new applications that require integrated, low-cost, and compact WINS technology. Micropower WINS system-on-a-chip technology has created a new, embedded, and densely distributed computing and sensing platform.

4.5 Plug-and-Play Sensors and Networks

Communications with sensors had been limited to either wired connections between the sensing elements and the data acquisition electronics or proprietary wireless communication protocols. Today wireless communication is common practice and some systems extend to the Internet by using wireless connections to sensor nodes via suitable protocols such as Bluetooth. One way of achieving wireless networks is by the employment of plug-and-play devices.

Plug-and-play is a term used in the computer field to describe a computer's ability to have new devices, normally peripherals, added to it without having to restart the computer. There are a number of terms or variations that describe similar abilities, including PnP and hot swapping. The term plug-and-play is most associated with Microsoft, who started using it in reference to a series of Windows 95 products. Plug-and-play enables users to change components and settings on a computer without having to know a great deal about those settings, because the computer recognizes the optimal configuration for the device and adjusts those settings accordingly.

There are many different types of plug-and-play sensors and devices available in the marketplace that can be used for a wide range of applications such as:

- Speed-critical applications requiring fast and easy setup—new product testing and development.
- Temporary applications requiring frequent reconfiguration—testing new products.
- Applications that make it impossible to use wires—health monitoring and other dynamic systems.
- Applications in harsh environments—environments with excessive heat, cold, humidity, corrosion, or radioactive conditions.
- Applications where cabling costs too much—remote operations.
- Mobile applications—military applications.

Two important factors in the wireless sensor networking are the network configuration and network management. In wireless networks, if a central controller is involved, the controller must know which sensors are connected and when, which sensors are within range of the communication link, and if there are new sensors added to the network or taken away. It may be difficult to keep track of the network configuration if the configuration changes continuously, as in the case of mobile sensors. Plug-and-play sensors address some of these network problems by defining an architecture with a standardized physical interface to the network, thus allowing a wide variety of sensor types to be connected on the same network, and defining a self-identification protocol, allowing the network nodes to describe and configure themselves dynamically to the network.

Self identification and self-connection features employed in plug-and-play sensors are achieved by the use of SIO. SIO converts various sensor outputs into standard high-level analog voltages for communication to the network nodes. The node that contains an SIO converts analog signal into a suitable digital form to be transmitted over a wireless link to the host computer. By providing this normalized output along with information describing the sensor and associated signal conditioning, communication between the network node and the host computer can occur in a user-selected manner. This process significantly improves connectivity and simplifies the information management process.

Two factors have promoted widespread adoption of plug-and-play sensors: the IEEE P1451.4 smart transducer interface standard and the Internet. IEEE P1451.4 is a standard for self-describing analog sensors using standardized TEDS, as explained in Section 3.5.8. The Internet brings the plug-and-play concept to legacy sensors and systems via distribution of so-called virtual TEDS. The new generation of measurement and automation systems uses these concepts to become robust, convenient to networks, and much smarter.

Typical examples of plug-and-play sensors are the MTS400/420 series of multifunctional boards offered by Crossbow Technology, Inc. (www.xbow.com/). These boards incorporate sensors for measuring temperature, humidity, barometric pressure, light, acceleration, and so on. The MTS400/420 series environmental monitoring boards combine processor/radio boards, thus allowing the creation of a wireless sensor network for environmental monitoring. These boards operate at 2.4 GHz frequency, complying with IEEE/ZigBee standards. Applications for the MTS400CA/420CA range from a simple wireless weather station to full mesh networks of environmental monitoring nodes.

4.5.1 Bluetooth and Plug-and-Play Sensor Networks

Bluetooth is becoming a widely applied worldwide standard for low-cost, short-distance local wireless communication operating in the license-free 2.4 GHz band. Bluetooth is extensively applied in laptops, handheld computers, mobile phones, PCs, and a wide range of other devices.

There are many companies offering Bluetooth products suitable for consumer and industrial applications. Much of the system architecture based on Bluetooth integrates hardware and software together to enable wireless sensors to be used in consumer products as well as industry. By leveraging Bluetooth wireless technology, dozens or even hundreds of sensors can be networked in a particular application. The Bluetooth-based networks can link different brands and types of sensors and data acquisition systems, such as handheld devices, as well as laptop computers and desktops to the Internet environment, as shown in Figure 4.17.

Since Bluetooth is a global standard and uses the universally available, unlicensed 2.4 GHz RF spectrum, Bluetooth-certified devices operate in the same way anywhere in the world. For example, a Bluetooth-certified PDA works with any PC equipped with a Bluetooth-certified card, regardless of the manufacturer.

The Bluetooth Special Interest Group (SIG) is driving development and application of Bluetooth technology. The Bluetooth SIG includes promoter companies such as 3Com, Ericson, IBM, Intel, Lucent, Microsoft, Motorola, Nokia, and Toshiba (more than 2000 companies). It is estimated that there are more than 700 million Bluetooth devices currently in use worldwide.

In typical plug-and-play applications, the user mounts sensors on a machine or in a facility (e.g., buildings) at a number of points to be monitored. The individual sensors are then connected to a wireless node. Any number of nodes can be mounted in an application area. Each node controls up to four sensors and incorporates a Bluetooth radio for wireless communication with a computer or handheld device. The node collects data from multiple sensors and transmits the data via Bluetooth wireless communica-

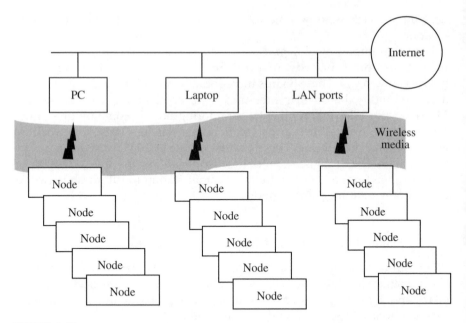

FIGURE 4.17
Wireless sensor networks via the Internet.

tion technology to a network hub or other Ethernet appliance such as a computer. Nodes can accommodate analog or digital signals.

There are many Bluetooth-based wireless plug-and-play devices for PCs and laptop computers, such as wireless serial RS-232 interfaces, USBs, PCM-CIA/PC cards, Fax, and modems. Serial wireless RS-232 interfaces are widely used in industrial, medical, and automotive applications for interfacing sensors, measurement devices, robotics, modems, and scanners. A typical example of a Bluetooth-based plug-and-play system is the Garnet M, by TADLYS Ltd. (http://www.tadlys.com/). This system is used for information gathering from medical equipment in hospitals and clinics.

4.6 Industrial Wireless Networks and Automation

Network topologies and supporting electronic components for industrial wireless networks are similar to wireless sensor networks and WINS, as explained in Section 4.3 and Section 4.4. Therefore details on components and network topologies will not be repeated in this section. However, industrial wireless sensor networks are in a transitional state as they are adapting to existing installations in factories and production plants. Many industrial operations contain fieldbuses, and wireless sensors need to be integrated with the existing instrumentation networks. In this section, adaptation of wireless systems to existing industrial infrastructures and infrastructure independent wireless networks (e.g., industrial mesh networks) will be discussed and some examples given.

4.6.1 Wireless Fieldbuses

Many fieldbus vendors are aware of the potential benefits of wireless technology, thus encouraging integration of wireless systems in their networks. There are numerous applications where wireless products are offering solutions; for example, when the configuration of control equipment requires information from a moving object or where alterations in wiring are difficult and expensive.

Popular wireless technologies for industrial automation are DSSS and FHSS. Spread spectrum reduces the influence of electrical noise and intentional jamming, and it matches receivers to enable messages to get through. It also produces a noise-like spectrum that reduces the chance of signals interfering with each other. This answers the concern that the many new unlicensed wireless networks may interfere with or even override existing licensed wireless networks. DSSS and FHSS ensures reliable communication in hostile RF environments. They handle multipath communication and offers better resistance to physical objects in the communication path that might cause interference.

All the necessary background information on fieldbuses (Section 1.6) and wireless systems (Chapter 2 through Chapter 4) necessary for wireless fieldbuses are given in this book. The only thing that can be added is that industrial wireless components and systems need to be rugged and robust for industrial applications. Here, repetition will be avoided, but some typical examples will be provided.

Wireless industrial fieldbuses are available for a range of products; some examples are available from Industrial Networking Solutions (http:// www.industrialnetworking.com/). The products of this company include wireless/IP Ethernets and wireless fieldbus gateways. For example, the SEM2411 is a high-speed wireless Ethernet bridge featuring 1.23 Mbps data transfer up to a 2.5 km communication range. It has up to 1 Mbps throughput operating at 2.4 GHz using FHSS technology. This bridge is designed for networking Ethernet compatible devices such as sensors, PLCs, and computers. Similarly, the BAT 11B wireless Ethernet bridge uses 2.4 GHz DSSS, complying with IEEE 802.11b standards. Also, the 905U-G wireless gateway provides a wireless interface between various fieldbuses used in process and automation applications. It supports fieldbus protocols such as Profibus, Ethernet, Modbus, DF1, DeviceNet, and Modbus Plus. The 905U-G includes an integral 900 MHz license-free radio transceiver, and transfers transducer and control signals (I/O) using a secure and reliable radio protocol. Other features of the 905U-G are high-security data encryption, automatic acknowledgment and error correction, peer-to-peer addressing, and multipath routing.

Among many other companies, Grid Connect (http://www.industrialeth-ernet.com/) has a range of industrial wireless network hardware components and software suitable for Bluetooth, Ethernet, 802.11b, serial, and USB products for short- and long-range operation.

4.6.2 Wireless Industrial Mesh Networks

Mesh networks are mostly based on cellular phone style radio links, using point-to-point or point-to-multipoint transmission. However, studies indicate that traditional wireless networks have liabilities in industrial applications. These include rigid structure, meticulous planning, and dropped signals during the communication. In contrast to the cellular phone-type mesh structure, wireless ad hoc mesh networks also proves to be effective. An advantage of ad hoc networks is that they are multihop systems. In multihop systems a node can communicate with a number of nodes in the neighborhood, thus they assist each other in transmitting packets through the network. This is particularly important if the operating conditions are hostile for RF transmission, as is the case of industrial operations, where heavy machinery and high voltage power exists.

An ad hoc multihop network is an example of a wireless mesh network. This particular system was first developed for industrial monitoring and control applications. It was built to communicate in point-to-point mode or peer-to-peer mode. In ad hoc networks, a node in the mesh can send and receive messages and can also function as a router that can relay messages for its neighbors. Through the relaying process, a packet of wireless data can find its way to the destination by passing through a number of nodes, as shown in Figure 4.18. The message relaying process in ad hoc networks

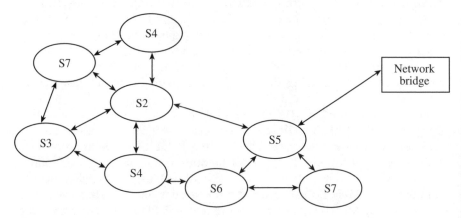

FIGURE 4.18
A wireless mesh network.

forms reliable communication links and is used in many applications. Also, ad hoc networks can operate with minimal preparation and planning, and they provide reliable, flexible networks that can be extended to thousands of devices.

In general, the mesh topology enhances the overall reliability of the network, which is particularly important in most industrial environments. Similar to Internet and other peer-to-peer router-based networks, a mesh network offers multiple redundant communication paths through the network. If one link fails for any reason, the network automatically routs the message through an alternate path. Another advantage of mesh networks is that they can be installed in a short period of time compared to their wired counterparts. These networks do not require sophisticated planning and on-site mapping to achieve a reliable communication.

One use of a mesh network is in distributed industrial control systems. There has been considerable effort expended in recent years to place more intelligence throughout control systems. The IEEE 1451 standard is evidence of this. Distributed intelligence is naturally served by wireless multihop mesh networks. Control of the wireless system is distributed throughout the network, allowing intelligent peers to communicate directly with other points on the network without having to be routed through a central control point.

Wireless mesh communication with distribute control facilities provides a whole new dimension in interactions between sensors and sensor clusters. Sensors can communicate directly with other devices on the network. This makes calibration and troubleshooting much easier than with modular systems. Modular systems enable localized decision making.

Point-to-point links, also known as a wireless bridges, replace a single communication cable. They may be used in many applications, for example, for connecting PLCs to remote monitoring stations. Point-to-point links can communicate reliably as long as the two end points are close enough to one another to escape the effects of RF interference and path loss. A reliable connection can be achieved by relocating the radios or boosting the transmit power to achieve the desired reliability.

Point-to-multipoint links are wireless systems (e.g., IEEE 802.11 or Bluetooth) that have one base station or access point that controls communication with all of the other nodes in the network, also referred to as hub or star topology. In this arrangement, point-to-multipoint networks converge at a single access point. The reliability of these networks is set by the quality of the RF link between the central access point and the end points.

In industrial settings, it can be hard to find a location for an access point that provides dependable communication with each end point. Moving an access point to improve communication with one end point may degrade communication with other end points. A wireless mesh network does not need a system administrator to control it. Getting messages to the destination is done automatically. In these networks, multiple nodes cooperate to relay a message to the intended destination. The mesh network is self-organized and does not require manual configuration. Because of this adding, new nodes or relocating existing nodes is done automatically.

Reliability and adaptability are the most important attributes of a wireless network for industrial control and sensing applications. Point-to-point networks can provide reliability since the devices in the network are not required to handle more than one or two end points, thus communication algorithms can be tailored for reliability and adaptability. On the other hand, point-to-multipoint networks can handle more end points, but their reliability may deteriorate due to task identification of the nodes and multiple reception of information coming from different nodes. Also, the position and placement of the access points and end points can affect the reliability and adaptability.

If environmental conditions result in poor reliability, it may be difficult or impossible to adapt a point-to-multipoint network and increase reliability at the same time. In contrast, mesh networks are inherently reliable and can adapt easily to environmental or architectural constraints. It is also possible to scale them to handle many end points at the same time.

An important goal in the operation of a mesh network is adaptability. Wireless mesh networks emulate Internet operations; for example, if one router goes down on the Internet, messages are sent through alternate paths by other routers. Similarly, if a device or its link in a mesh network fails, messages are sent to the destination via other devices. Therefore the loss of one or more nodes does not necessarily affect network operation. A mesh network is self-healing because human intervention is not necessary for rerouting of messages. Also, a weak signal or dead zone can be fixed simply by dropping a repeater node in place.

The meaning of redundancy in the real word is a matter of degree and must be carefully specified. In a mesh network, the degree of redundancy is essentially a function of node density. A network can be deliberately overdesigned for reliability simply by adding extra nodes, so each device has two or more paths for sending and receiving data.

A mesh network is also scalable and can handle hundreds or thousands of nodes. Since the network's operation does not depend on a central control point, adding multiple data collection points or gateways is possible. The distance between nodes can be shortened, which increases the link quality. If the distance is reduced by a factor of two, the resulting signal becomes at least four times more powerful at the receiver. This makes links more reliable without increasing transmitter power in individual nodes. General reliability, redundancy, and reach can be improved by adding more nodes in the network.

Industrial systems can benefit from wireless formats that satisfy the multiple conflicting demands of redundancy, distributed communication, flexibility, and reliability. In addition, self-configuring, self-healing networks are inherently less expensive to install and maintain, especially as radios and microprocessors become cheaper.

An application for wireless, multihop mesh networks is in the diagnostic monitoring of devices. Operations and sensor signals can be monitored for abnormalities without affecting the sensors' routine operations. If an abnormal signal or trend is observed, an alert is triggered. This alert is independent of normal operations.

Examples of wireless industrial mesh networks are offered by Expert Monitoring (http://www.expertmon.com/). This company produces a wide range of hardware and software for wireless networking in industrial environments. An unlimited number of transmitter modules can be integrated in a network that can be spread around a plant or site. Bluetooth technology is used for sending and collecting sensor data to a wireless network controller unit, which could be a PC. This wireless network runs on WiSView software.

Another example of a wireless mesh network is a case study conducted by Ember Corporation (www.ember.com/) to validate wireless mesh networks in challenging industrial environments. Researchers in this company designed and implemented a wireless control system in a water treatment plant. The environment contained significant hurdles for healthy wireless communication. The plant contained thick reinforced concrete walls, large metal objects, heavy machinery, high-current cables, and so on. The goal of this case study was to connect the instruments in the pipe gallery back to the control panel located in a remote control room. In order to achieve this, a computer was equipped with a wireless networking card connected to its serial port. Each process instrument also had bus connections replaced with wireless networking cards. A total of 12 instruments contained wireless cards. The locations of the wireless instruments included spots where RF signals had to pass through reinforced concrete walls. After some trial and error, it was shown that the RF links could be significantly improved by adding several relay and repeater points at carefully selected points. It was reported that the system operated successfully and there was no discernable difference between the wireless communication network and the wired serial cable network.

4.7 Conclusion

In this chapter, all knowledge gained in Chapter 1 through Chapter 3 was brought together. It was shown that modern wireless sensor and instrument networks can be obtained by using embedded or modular designs. The networks can be expanded using bridges, routers, and repeaters. The construction of wireless sensors and instruments was discussed and many examples provided. Power issues of wireless networks were also addressed. Two examples, wireless sensor networks and WINS, were discussed in detail. Applications of Bluetooth and IEEE 802 technologies were demonstrated. It was pointed out that wireless networks can be configured by point-to-point or point-to-multipoint communication methods.

5

Wireless Sensor and Instrument Applications

Previous chapters concentrated on information necessary for the construction and operation of wireless sensors and instruments. Important issues on networking have been discussed and examples of wireless sensor and instrument networks provided. Recent advances in communication and integrated circuit (IC) technology, and the emergence of wireless standards has resulted in the proliferation of wireless sensors and instruments. As a result, today, wireless sensors and instruments find many applications in industrial plants, consumer goods, scientific laboratories, test facilities, the military, aviation, homes and buildings, hospitals, networking, and so on. These application areas are growing consistently as vendors of wireless sensors and instruments respond to consumer demand by offering a diverse range of wireless devices. In this chapter, examples of wireless sensors, instruments and networks are given in the following areas:

- Application-specific wireless sensors and instruments.
- Commercial wireless stand-alone sensors and instruments.
- Research and development (R&D) in wireless networks.
- Industrial applications.
- Human health and environmental applications.
- Radio frequency identification (RFID).
- Consumer products and other applications.

This list is by no means exhaustive, as wireless instruments find many new and novel applications almost daily.

5.1 Application-Specific Wireless Sensors and Instruments

Wireless instruments have existed for decades, but they were expensive and mainly custom built for special purposes. In addition, wireless networks mainly existed for data transmission, particularly from remote areas in telemetric and supervisory control and data acquisition (SCADA)-type applications. Today, wireless instruments and networks are used in both large scale and local environments, basically with the potential of replacing all forms of wired connections in factories and plants, homes and buildings, laboratories, automation and testing, and so on. In this respect, wireless sensors, instruments, and networks are relatively new, but they are emerging in a cost-effective and competitive alternative to wired systems.

Although wireless instruments and networks are becoming commonplace, there are many specific situations for which the required instrument may not be available. This may be because manufacturers simply do not offer such products. In many cases, it may be necessary to construct a wireless instrument starting from basic principles and components. The information provided in this book can be applied to achieve the goal of constructing wireless instruments and networks. In this section, further guidance is provided.

Application-specific wireless instrument and network designers must be aware of a number of issues during the design, construction, and implementation process. Common components of wireless instruments include ICs or board mounts, circuit boards, panel or chassis mounts, modular bay or slot systems, rack mounts, DIN rails, and stand-alone systems. Common signal inputs available for instruments include DC voltage, DC current, AC voltage, AC current, frequencies, and charges. The sensor inputs include accelerometers, thermocouples, thermistors, resistive temperature detectors (RTDs), strain gauges or bridges, and so on. In addition, there may be specialized inputs that include encoders, counters, tachometers, timers, relays, or switches. Many products have integral sensors or transducers in IC form. Common outputs from the instruments include voltage, current, frequency, timer or counter, relay, resistance, and potentiometer.

Interfaces necessary for instruments include no display, front panel display, touch screens, handheld or remote programmers, and computer programmables. Host connection choices include direct backplane interface, RS-232, RS-422, ST485, universal serial bus (USB), IEEE 1394, general purpose interface bus (GPIB), small computer system interface (SCSI), transistor-transistor logic (TTL), parallel, Ethernet, modem, and radio or telemetry. The transmission rate of data is important to consider as well. Common applications for instruments include general laboratory, industry, environmental, vehicular, marine, aerospace or military, seismic or geotechnical, weather or meteorology, and medical or biomedical. Additional specifications to consider for instruments include application software, memory and storage, network specifications, analog-to-digital (A/D) conversion, filter specifications, amplifier specifications, and environmental parameters.

In the design of application-specific wireless instruments, some of the critical considerations include

- Level of integration—single-chip solutions, multichip modules, or ultracompact printed circuit boards.
- Power mode—the environment, batteries, mains, or other sources.
- Standards to adopt for communication—Bluetooth, IEEE 802.11, or other license free and industrial, scientific, and medical (ISM) bands, or others.
- Network standards and topologies.
- Security issues.
- Integration of the instrumentation system with the network of the organization.
- Management issues and solutions.
- Engaging other parties such as radio designers.

The construction of wireless sensors and instruments is discussed in Chapter 4. A number of methods can be used for designing and constructing application-specific wireless instruments and networks, including

- Wireless sensor systems that incorporate wireless sensors.
- Embedded instrument systems that incorporate wireless features within the sensor or instrument itself.
- Modular or add-on systems that convert wired sensors into seamless wireless connections via external or plug-and-play wireless equipment supplied by other parties.

5.1.1 Application-Specific Wireless Sensors and Networks

Sensor networks combine distributed sensing, computation, storage, and wireless multihop communication. Wireless sensor networks are used in a variety of scientific, military and commercial applications. In recent years numerous advancements have been made allowing networks of inexpensive sensor nodes to be deployed. The vast potential for sensor networks has been demonstrated by numerous scientific and commercial applications.

Sensors are integrated with miniature transmitters, receivers, and transceiver hybrids, plus original equipment manufacturer (OEM) data radio boards to produce wireless sensors. Related protocol firmware and software supports the operation of wireless sensors in a network environment. Wireless sensors are used to implement short-range wireless data communication and control links in many applications such as wireless barcode readers and label printers, credit card readers and receipt printers, automatic utility meter reading systems, sports and recreational equipment, industrial telemetry, RFID tags and access control badges, medical telemetry, and so on.

Wireless sensors require design at the basic IC level, and this is a highly specialized area on which many research organizations are concentrating.

Some examples of wireless sensors and networks are provided throughout this chapter. However, because of IC sensors, wireless instruments can be obtained by the using off-the-shelf components, as explained in the next section.

5.1.2 Application-Specific Embedded Wireless Instruments and Networks

Embedded wireless instruments contain all needed components, such as sensors amplifiers, filters, signal conditioners, converters, microprocessors, and radio frequency (RF) components. A simplified block diagram of an embedded wireless instrument is shown in Figure 5.1. Necessary information on the operation of embedded instruments, their components, and their construction is provided in Chapter 1 and Chapter 4. One or two further examples of some essential components for embedded wireless instruments suitable for application-specific systems are provided in this section as guidance to designers.

The 26PC series of pressure sensors from Honeywell are typical examples of sensors suitable for use on printed circuit assemblies to construct wireless instruments. The 26PC SMT supports gauge, vacuum gauge, differential, and wet/wet sensing applications. Although primarily designed for the medical industry, these pressure sensors can be used for industrial surface mount pressure sensor applications. They are temperature compensated and factory calibrated to operate in a range from 0 to 250 psi.

Analog Devices, Inc. (http://www.analog.com/) offers the AD8131 and AD8132 high-speed differential amplifiers for use in wireless communication, video, and instrumentation. The AD8131 has a fixed gain of 2400 MHz and a 2000 V/μs slew rate. The AD8132 has adjustable gain/feedback, 350 MHz bandwidth, and 1200 V/μs slew rate, and is designed for a variety of differential signal processing applications. Both devices have a –70 dB output balance error and are available in small outline integrated circuit (SOIC) and micro-SOIC packages.

The monolithic AD6644, from Analog Devices, can accurately convert wideband analog signals (200 MHz input bandwidth) for use in multichannel, multimode digital receivers (software radios); noise is 74 dB, sampling rate is 65 Msps, and distortion is 100 dB spurious-free dynamic range (SFDR). The device features less than 300 full-scale (FS) sampling jitter, differential analog inputs, and 2s complement digital output that is 3.3 V complementary

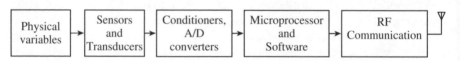

FIGURE 5.1
An embedded wireless instrument.

metal oxide semiconductor (CMOS) compatible. Applications include single-channel digital receivers for use in wide-channel bandwidth systems such as code division multiple access (CDMA) and wireless CDMA. The AD6644 comes in 52-pin packages.

The AD9226, from Analog Devices, Inc., is a 65 Msps analog-to-digital (A/D) converter suited for applications in cellular base station and software radio communication. The full-power input bandwidth is 750 MHz and the unit can digitize wideband signals with effective number of bits (ENOB) performance greater than 11 bits. Pin compatible with the AD922X family, the A/D converter features an on-chip sample-and-hold amplifier, clock duty cycle stabilizer, out-of-range indicator, and the ability to digitize a 70 MHz intermediate frequency (IF) signal with 86.4 dB. The device enables IF transceivers to be reprogrammed immediately to accommodate multiple modulation formats. Other applications include wireless local loop, radar, and high-speed instrumentation.

The MSP430 family of ultra low power 16-bit reduced instruction set computer (RISC) mixed-signal processors, from Texas Instruments, is suitable for battery-powered measurement applications. The power consumption is 0.1 µA random access memory (RAM) retention, 0.8 µA real-time clock modes, 250 µA/million instructions per second (MIPS) active. They include A/D converters, D/A converters, opamps, and a comparator. The company also offers development tools for MSP430 that include the TI MSP430 C compiler, in-circuit debuggers, development boards; and real-time operating systems (RTOS). The MSP430 family includes a flexible clock system with five low-power modes. The family features a typical standby current consumption as low as 0.8 µA with a real-time clock function active. Total power consumption is low due to fast instruction execution and it is able to start up from standby in less than 6 µs with a fully synchronized high-speed system clock.

Single-chip transceivers are used extensively in specific applications. An example is illustrated in Figure 5.2. Among many others, Texas Instruments offers a portfolio of radiofrequency integrated circuits targeted to transmit and receive data using ISM bands. Transmitter-receive products cover frequencies between 315 to 950 MH_2 supporting up to 64 kbps.

The CC1010, by Chipcon (http://www.chipcon.com/) is a complete RF system-on-a-chip. This device integrates a low-power 300 MHz to 1000 MHz RF transceiver with a data rate of up to 76.8 kBit/s and 8051-compatible microcontroller. The microcontroller has 32 kb in-system programmable flash memory, hardware data encryption standard (DES) encryption/decryption, and a three-channel, 10-bit A/D converter. This creates an embedded system with wireless communication and sensor interfacing capabilities. This device also has four timers, two pulse width modulators, two universal asynchronous receiver-transmitters (UARTs), real-time clock, Watchdog, and 26 general input/output (I/O) pins. It complies with EN 300 220 and FCC CFR47 Part 15 rules.

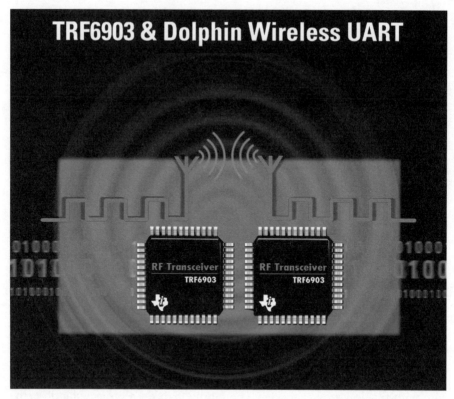

FIGURE 5.2
Typical example of a transceiver. (Courtesy of Texas Instruments: http://www.chipcon.com)

Another example of a single-chip transceiver is the MAX82XX family of products, by Maxim Integrated Products (http://www.maxim-ic.com/). MAX2828 and MAX2829 are single- and dual-band 802.11a/b/g transceiver ICs. MAX2825, MAX2826, and MAX2827 are 2.4 GHz/5 GHz, single- and dual-band 802.11g/a transceiver ICs. They are based on proprietary, second-generation silicon germanium (SiGe) bipolar CMOS (BiCMOS) process. The receiver has a noise figure of 3.5 dB at 2.4 GHz and 4.5 dB at 5.2 GHz. These devices are suitable for applications in wireless fidelity (Wi-Fi) network interface cards (NICs), access points, wireless routers, Wi-Fi mini-peripheral component interconnect (PCI) cards for notebooks, and Wi-Fi modules for portable handhelds.

The PKLR2400S-100, from AeroComm, Inc. (http://www.aerocomm.com) is a 1.65 in. × 2.56 in. × 0.20 in. shielded, ready-to-embed, frequency hopping, spread spectrum transceiver designed for integration into any industrial system operating in the unlicensed 2.4 GHz band. The range of transmission is up to 160 m indoors and more than 3.3 km line-of-sight. The unit provides an asynchronous TTL-level serial interface for OEM host communication, including system and configuration data, and operates in point-to-point or multipoint client/server architectures. The device features proprietary ConnexSync technology that integrates a custom IC with tightly connected transmit/receive functionality, which provides for faster synchronization schemes, resulting in high data transfer rates. Also, 900 MHz and 848 MHz modules are available.

Applied Wireless (http://www.applied-wireless.com/) offers a line of 900 MHz transmit and receive modules for data transmission with ranges up to 350 m with unity gain antennas. Ranges up to 1.5 km are achievable with a high-gain Yagi antenna on the receiver. Modules comply with FCC Part 15 unlicensed requirements. The FCC Part 15 rule states that for continuous or regular transmission intervals, the 900 MHz band must be used. The 900 MHz band can be used for periodic data transmissions as well. The product line includes the T900FM (transmitter) and R900FM (receiver) for medium-speed data transmission with data rates of 10 kbps to 200 kbps. These units are surface mountable, operate on a 3 V power supply, and are suited for use in remote control, security, RFID, telemetry, video, and data transmission applications. They also have 300/400 MHz modules, but FCC Part 15 requirements limit 300/400 MHz frequencies to remote control-type applications.

There are thousands of examples of similar components, suitable building blocks of wireless instruments, some of which were mentioned in Chapter 1 and Chapter 4.

5.1.3 Application-Specific Modular and Add-On Wireless Instruments and Networks

Modular and add-on systems are based on the integration of stand-alone modules, as illustrated in Figure 5.3. In this case, conventional instruments with analog signal or digital data output are connected to a wireless communication device or RF modules. Many of the commercially available RF modules accept multiple analog or digital inputs and provide RF output. RF modules are designed to accept different forms of analog inputs, such as 4–20 mA, 0–5 VDC, 0–10 VDC, and so on. In some cases, transceivers are integrated with the instrument using an intermediate device to make the

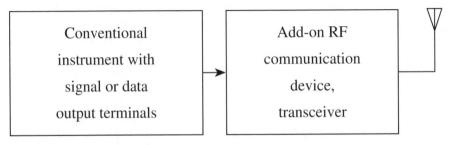

FIGURE 5.3
An add-on wireless instrument.

instrument's signals acceptable for RF transmission. Some application-specific cases include data loggers, SCADA, oil wells, water wells, industrial plants, processing plants, power plants, gas plants, flow rates, water level, voltage, current, temperature, high and low pressure, strain, liquid level, inclinometers, gas detection, and accelerometers.

Radio frequency modules and add-on devices come in a variety of forms operating at different frequencies. These frequencies may be in the license-free ISM or other bands. In today's applications, two standards are of particular importance: Bluetooth and IEEE 802.11. The Bluetooth standards are explained in Chapter 3 and Chapter 4. Because of its growing importance, IEEE 802.11 will be revisited with a particular emphasis on application-specific implementations.

IEEE 802.11-based networks and components are designed to operate in the unlicensed 2.4 GHz ISM band. Federal Communications Commission (FCC) rules dictate the use of either frequency hopping spread spectrum (FHSS) or direct sequence spread spectrum (DSSS) technology. FHSS systems are targeted at low-cost, low-power, short-range, and low data rate applications. The applications of DSSS and FHSS technologies include Bluetooth, HomeRF (SWAP), 2.4 GHz digital European cordless telephone (DECT), and IEEE 802.11. The DSSS IEEE 802.11b systems are intended for higher data rate and longer range applications, and typically consume more power. IEEE 802.11 FHSS systems currently support data rates up to 1.6 Mbps, but at a new 5 MHz channel bandwidth allocation. The wideband frequency hopping allows data rates up to 10 Mbps. This will allow FHSS to compete with the current 11 Mbps data rate supported by DSSS systems.

For FHSS, the 2.40 GHz to 2.835 GHz ISM band is broken into seventy-five 1 MHz channels, with a 2 MHz lower guard band and a 3.5 MHz upper guard band. All FHSS systems are based on time division multiple access (TDMA), with the number of frequency hops per second varying from one standard to another. Bluetooth and DECT utilize Gaussian frequency shift keying (GFSK) modulation, whereas HomeRF and FHSS 802.11 use two-level and four-level frequency shift keying (FSK) to take advantage of the higher efficiencies offered by saturated power amplifiers.

There are many single-RF IC transceiver solutions for FHSS wireless systems. Many transceivers contain low-noise amplifiers, power amplifiers, power amplifier predrivers, voltage controlled oscillators (VCOs), and frequency multipliers integrated in the chip. Often power amplifiers integrate a power detector and closed-loop power control to provide nearly constant output power over the full range of supply voltage, temperature, and input power levels. Some power amplifiers feature analog gain control, integrated input matching, and a low-power shutdown mode. VCOs containing additional circuits, such as an integrated tank circuit and on-chip varactors, provide a linear modulation input, optimizing it for direct frequency modulation applications. In some cases, frequency multipliers following the VCO convert the low frequency to the 2.4 GHz range.

Miniature data transmitters are typical examples of modules that can be added to existing instruments to make them wireless. For example, the DL160 series RF transmitters from APT Instruments (http://www.aptinstruments.com/) are battery-powered 54 mm × 35 mm × 15 mm devices that transmit data to a receiving antenna attached to the serial port of a PC; resolution is 16-bit and the operating frequency is 418 MHz at a range of about 250 m. RF transmitters provide wireless transfer of monitoring and control signals between sensors, monitors, controllers, and data collection devices. Wireless systems simplify installations in hard to reach locations and eliminate ground loop problems. The company offers seven models; sensors are either internal or connected through screw terminals. Time-sampled and date-stamped data are stored on the PC. The logger can be set up on a PC with the included Windows-compatible software; National Institute of Standards and Technology (NIST)-traceable calibration is available on most models.

Quatech (http://www.quatech.com/) offers the QTM-8524 DSSS RF modem module for its QTM-8000 series of remote data acquisition and signal conditioning products. The modem modules use Gaussian minimum shift keying (GMSK) modulation, operate in the 2.4 GHz ISM band, and convert RF signals to serial RS-232 or RS-485 for communication with a PC or serial network. The data acquisition module's standard package includes either an indoor/outdoor wall-mount swivel antenna or a ceiling-mount antenna. These devices are suitable for networking PCs and programmable logic controllers (PLCs). QTMSuite software contains a Windows-based utility for configuring I/O parameters and sending commands through the network. At least two QTM-8524 modules are required to implement a wireless system. For more complex applications, QTMSuite supplies a server, 32-bit Windows dynamic link libraries (DLLs), ActiveX OCX controls, a custom set of LabView virtual instruments, and a custom DASYLab. The dual watchdog protection minimizes possible damage if a portion of the network fails. The system has the ability to simultaneously implement multiple modules operating at different baud rates with different data formats and at different frequencies. Power for controlling the entire system can be implemented by a PC via its standard serial port.

Series TR300 transceivers, from Otek Corp. (http://www.otekcorp.com/), are constructed for plug-and-play applications of PC-to-PC wireless communication. Typically the TR300 receives data through its serial port (RS-232/485/TTL), stores it (up to 72 characters), and automatically converts and transmits it at 19.2 kbaud or 115.2 kbaud (on-off keying [OOK] or amplitude shift keying [ASK] modulation in the ISM band) to the other wireless end, where it is received and converted back to serial. Peak transmit power is 1 mW. Up to 15 nodes, plus a master, can be addressed using the built-in address switch. The devices are supplied with PC software, instructions, and mounting guides.

Original equipment manufacturer (OEM) is a company that manufactures a product and sells it to a reseller. Maxstream (http://www.maxstream.net/) products are an example of such devices. The 9XTend of the Maxstream RF module provides 256-bit advanced encryption standard (AES) security. It is able to communicate peer to peer, point to point, and point to multipoint. It uses FHSS with 10 hopping channels, each with more than 65,000 possible network addresses with three network filtration layers. Host interface baud rates vary from 1200 bps to 230,400 bps. Multiple low-power modes, including shutdown pin, cyclic sleep, and serial port sleep, provide a current consumption of 1 μA. This device communicates to a range of up to 64 km line-of-sight at a data rate of 9600 baud, up to 32 km line-of-sight at 115,200 baud RF, up to 900 m in indoor/urban environments at 9600 baud RF, and up to 450 m in indoor/urban environments at 115,200 baud. It operates on a 2.8 V to 5.5 V power supply with 1 W power output. The power output is software selectable in the range of 1 mW to 1 W. The receiver sensitivity is –110 dBm at 9600 baud.

Portable radio modems, illustrated in Figure 5.4, represent a range of products that can be used for wireless instrument networks. A networked radio modem with an internal nickel metal hydride (NiMH) battery is the IC-100MES, from Monicor Electronic Corp. (www.monicor.com). The modem operates on narrowband FM (450–470 MHz) with a transmitter output power of 250 mW. The range of the modem is up to 1.6 km line-of-sight using an internal stubby antenna. External antennas enable greater ranges. The RS-232 interface has selectable baud rates; individual nodes

FIGURE 5.4
A portable radio modem. (Courtesy of Monicer Electronic Corp., http://www.monicor.com/ic100mes.html)

transfer American Standard Code for Information Interchange (ASCII) data, control codes, and protocol bits at a maximum speed of 4800 bps. The data rate between the base radio station and a host computer is up to 19.2 kbaud and it uses 16-bit error detection and correction techniques. The communication protocol is a scan sequence with collision detection or it is multidrop broadcast RS-422/485 compatible. Up to 99 nodes can be addressed for portable and laptop computers and peripherals operating in point-to-point, networked, or multidrop modes.

The Wave-Lynk CT-4000 series of license-free 2.4 GHz FHSS data and sensor interface systems from Wave-Link Wireless, Inc. (http://www.wave-lynk.com/) enable high-speed wireless data logging for heavy-duty industrial and sensing applications. The units can transmit data 16 km line-of-sight to a host/base station. The NEMA 4X- and IP66-rated (here IP stands for Ingress Progress) devices are available in 10 mW and 100 mW versions, with RS-232 and high-impedance sensor bridge interfaces. Custom versions, including battery-powered embedded systems, are available.

The Copy Cat digital system, from Omnispread Communications, Inc., (http://www.omnispread.com/), utilizes the spread spectrum and packet radio technologies and operates the radio in the 915 MHz ISM band. Each Copy Cat digital system is uniquely preset in the factory so that it works as a plug-and-play system. This packet radio system requires air time of 10 ms to execute the switching command, reducing the probability of interference. If interference occurs, the system retries until the switching command is executed correctly. This company offers other similar products for RF networks, such as the LAWNII+ starter kit, to set up a point-to-point network for a distance up to 4.6 km. It includes two LAWNII+ radios, two 8.5 dB Yagi antennas, low-loss RF cables, data interface cables with the power supply, two aluminum raincoats, LAWN 3.0 terminal emulator software, and a manual.

There are many other examples of such devices that are suitable for producing application-specific wireless instruments for networking purposes.

5.2 Commercial Wireless Sensors and Instruments

In the beginning of this decade, wireless instruments and networks existed but were scarce and expensive, offered only by a handful number of companies. Today, many companies are offering value-added wireless solutions for specific markets. Some integrate sensors and radios to reduce system cost. With spread spectrum radios now available for less than $20, applications once thought out of reach are now being reevaluated. Thousands of wireless instruments now exist in the marketplace. In this section, further examples of wireless sensors and instruments are provided. Some examples have already been given in Chapter 4.

One important development in the use of technology is that narrowband radios are disappearing and are being replaced by broadband and ultra wideband systems using spread spectrum techniques. Narrowband systems are suitable for use in harsh environments. The newest wireless products on the market are almost all spread spectrum, and the contest between direct sequence and frequency hopping continues. The newest contender, ultra wideband, is making inroads in the marketplace, but it is still too early to comment on its success. However, ultra wideband is likely to be a serious option in many applications in the coming years. Some examples of commercially available wireless sensors and instruments are given below.

TorqueTrak 9000, from Binsfeld Engineering Inc. (http://www.binsfeld.com/), uses a miniature battery-powered transmitter that can be mounted on a rotating shaft and broadcasts digital data from a torque-sensitive strain gauge, converting the shaft parameters into a rotating torque sensor. The transmitter frequency is 903-922 MHz with a transmission distance of about 10 m. This system is a microprocessor-based system that features 7 user-selectable gain settings, 8 user-selectable broadcast frequencies, auto-zero, and a low battery indicator. The receiver can receive signals from 8 different transmitters. System resolution is 14 bits, with 0–500 Hz frequency response.

Wireless sensor controllers are used to monitor and control sensors in industrial settings. For instance, the RMT10, from Advanced Embedded Systems (http://www.advancedembedded.com/) accepts a variety of standard sensor inputs and provides A/D conversion. It either stores data or transmits over a two-way wireless RS-232 connection. The wireless connection incorporates a Maxstream ConnexRF transceiver module, enabling point-to-multipoint communication in the license-free 2.4 GHz band. The transceiver employs FHSS and provides communication in indoor and outdoor operations. Features include two analog inputs, two analog outputs, a NEMA 4 enclosure, and a 2.4 GHz, channel hopping, packet data radio.

The WAU80/160, from Mastery Instruments (http://www.masteryinstruments.com/), is an 8- or 16-channel, 500 ksps, 12-bit D/A system that includes a wireless connection to the data collection PC. The wireless data link is a 2.4 GHz, frequency hopping, spread spectrum system that features automatic repeat request and forward error correction; real-time data rates are up to 5 ksps. It has a range of 3.3 km line of sight. A TF115 general purpose digital modem is required on the receiving site. An onboard buffer holds up to 48 MB; a LabVIEW driver is included. The enclosure is NEMA 4 rated.

Wireless weather stations find extensive professional and consumer applications. An example of a wireless weather station for outdoor and indoor use is shown in Figure 5.5. This wireless sensor monitors and displays outdoor and indoor temperature and humidity. Outdoor operation can support three sensors. Among other manufacturers, Oregon Scientific (http://www2.oregonscientific.com/) offers a range of weather station products.

FIGURE 5.5
A wireless weather station. (Courtesy of Oregon Scientific, http://www2.oregonscientific.com)

Another example of a wireless weather system is the Cell-Alert AWS-2100 wireless Internet telemetry weather station, by Strison Wireless Systems (http://www.strison.com/), sends digital data packets through an underused channel on existing cellular towers through North America. Designed to perform microclimate weather monitoring, the system has a solar power supply and monitors ambient temperature, relative humidity, wind speed and direction, rainfall, solar radiation intensity, and soil temperature. Measurement values are stored and transmitted at programmable intervals along with data for minimum, maximum, and summations, as well as calculated values. Alarm events are transmitted in real time. Transmitted data are received by the company's back office for storage and processing.

5.3   Wireless Instruments and Sensor Networks in Research and Development

There are many on-going research and development (R&D) projects that aim to improve the hardware, software, and firmware of wireless sensor and instrument networks. R&D efforts in instruments and sensor networks are being concentrated in three main areas: hardware and software aspects of device operations and the physical level, efficiency at the physical and network level, and communication protocols and network management of the system at both the physical and network level in compliance with the open system interconnection (OSI) reference model.

The majority of these R&D projects primarily aim to provide a service infrastructure or middleware that supports the operations of complex sensor and instrument networks. The classical approach to wireless sensor networks and to the supporting devices centers around low-power and highly integrated hardware nodes. One of the most prominent studies in hardware comes from the University of California, Berkeley. The mote wireless system has been commercialized and used by many researchers. Other prototype wireless systems are also available, typically based on commercially available popular hardware components as well as supporting software.

A study on the dynamic sensor networks based on BT nodes was reported by Beutel et al. (2004). This study concentrated on development of firmware and software and therefore will be reported in detail in the next subsection. The BT nodes study exemplifies trends in the R&D of wireless sensors and networks at the operational and physical level. Efficiency and power issues of the wireless sensors and networks are discussed in Section 5.3.2 and network management issues are addressed in Section 5.3.3.

5.3.1   Hardware and Software Issues at the Operational and Physical Level

There is much R&D effort concentrating on the hardware and software aspects of wireless sensors and instruments. Some of the R&D efforts concentrate at the chip level to design and construct stand-alone wireless sensors and instruments that can be produced cost effectively (e.g., S.Q. Wang et al., 2004), whereas other researchers are concentrating on the improvement and miniaturization of existing devices.

An example of R&D at the operational and physical level using an established standard is BTnodes. BTnodes are based on the Bluetooth wireless network technology, as it provides a rich reservoir of Bluetooth-enabled devices, actuators, infrastructure gateways, and user interfaces that can be used to configure novel systems. Bluetooth is an interoperable wireless networking standard that is implemented in a wide range of consumer devices such as personal digital assistants (PDAs), laptops, mobile phones, digital cameras, etc. It is also used as a hardware extension, such as USB, RS-232 ports, Personal Computer Memory Card International Association (PCM-CIA) cards, and so on. Interoperability with a wide variety of devices and easy configurability make Bluetooth technology a valuable tool to try in different and novel applications such as wireless sensor networks.

Sensors can be configured as single and stand-alone nodes backed up by suitable processors, wireless communication devices, and power supplies, as explained in Section 5.1. Large and dense networks of these nodes can be deployed to monitor a wide variety of real-world phenomenon or a distributed process. However, a typical sensor node can only provide limited information about the system, since it perceives only a small local subset of its environment. For this reason, the sensor nodes need to cooperate with one another to exchange information to compensate for each other's weaknesses and share valuable limited resources.

The basic system concepts of the study allow internode interaction between the sensors. This internode interaction is called distributed tuple space. For more computing power, outsourcing is required in which the sensor networks are connected to some back-end infrastructure for tasking the functions of the network nodes, data storage, and evaluation of the results. Generally, limited resources available at the sensor nodes preclude execution of intensive data processing; such services are usually backed up by external infrastructure with more computing and memory resources, such as the Internet or large base stations.

In a prototype application of BTnodes reported by Beutel et al. (2004), the hardware consisted of an Atmel Atmega128L microcontroller with on-chip memory and associated peripherals. The Atmel Atmega128L microcontroller features an eight-bit RISC core delivering up to 8 mega instructions per second (Mips) at a maximum of 8 MHz. On-chip memory consists of 128 kb of in-system programmable flash memory, 4 kb of static random access memory (SRAM), and 4 kb of electrically erasable programmable read-only memory (EEPROM). Other integrated components are software for debugging, timers, counters, pulse width modulation, 10-bit A/D converter, inter-IC (I^2C) bus, and two hardware UARTs. External low-power SRAM adds an additional 240 kb of data memory to the BTnode. An external quartz oscillator supports timing updates while the device is in low-power sleep mode. For the radio transmitter, an Ericsson Bluetooth module was connected to one of the serial ports of the microcontroller using a detachable module carrier and to a planar inverted F antenna that is integrated into the circuit board.

Network solutions with BTnodes have been implemented by the use of gateways, which are used to bridge the gap between the sensor networks and the Internet. Off-the-shelf Bluetooth devices are used for the networks and Bluetooth-enabled computers and laptops that act as a bridge between the network and wireless local area network (WLAN) was selected. It is worth mentioning at this point that Bluetooth-enabled mobile telephones are used for information transmission in some applications.

Algorithms and the development of and applications for these algorithms for sensor networks are a nontrivial task for a number of reasons. First, sensor networks are highly dynamic distributed systems. Second, the behavior of the sensor network is highly dependent on the physical environment. Third, for reasons of energy efficiency, networks perform in-network data processing and aggregation, thus minimizing the volume of data to be transmitted. To verify and debug such systems, developers need to have a good knowledge of raw sensor data and aggregate output of the sensor network.

Software in the BTnodes is made up of low-level drivers that are interrupt driven and a simple dispatcher for scheduling multiple threads. This operating system is well suited for the application of small-scale networking devices that consist of simple I/O and monitoring tasks with wireless communication capabilities. Using the operating system, event-driven programming can be developed so that smaller events can be handled sequentially. Large tasks can be tackled by breaking the tasks into smaller pieces, which can be treated as small events.

The Bluetooth software stack provides networking and driver functionalities. A subset of the Bluetooth specification is implemented and accessible to the application through an application programming interface. Bluetooth link management is performed on the L2CAP layer. RFCOMM, the serial port emulation of Bluetooth, provides connectivity to computer terminals and consumer devices such as cameras and mobile phones. Through RFCOMM, the BTnode can dial up an application server using a mobile phone or make use of other Global System for Mobile Communications (GSM) services.

In the BTnodes study, the system software is designed for portability and is available for different emulation environments (x86 and iPAQ Linux, Cygwin, and Mac OS X), apart from the embedded platform itself. Emulation simplifies application building and speeds up debugging. A BTnode system software kit consists of a building environment (avr-gcc cross-compiler and standard libraries), source code, debugging support, demonstration examples, and documentation.

Besides the core system software, the BTnode platform also provides a means of integrating sensor nodes into the surrounding communication infrastructure, to exchange and collaboratively process the data of the local distributed sensors.

BTnode system provides services for utilizing mobile phones as infrastructure gateways. An RFCOMM connection is set up between a BTnode and a nearby phone, over which standard advanced technology (AT) commands are sent to the phone using the serial port profile (SPP) and dialup network profile (DUN) of Bluetooth. AT commands manage and control GSM data connections using a background infrastructure server. Alternatively, a BTnode can embed sensory data into a short message service (SMS) message and transfer it.

Power consumption in wireless networks is a well-researched area. One of the shortfalls of Bluetooth technology is it requires long connection setup times and a lower degree of freedom with respect to possible network topologies. On the other hand, Bluetooth's connection-oriented features ensure interoperability between different devices, thus allowing standardized, trouble-free interfaces. Bluetooth offers a significantly higher bandwidth compared to many low-power radios. The standard host controller, human-computer interaction (HCI), provides a high-level interface that requires no knowledge of the underlying baseband and media access layers and their respective processing. This offers built-in high-level link layer functionalities such as synchronous and asynchronous communication links, multiplexing, integrated audio, forward error correction, automatic packet transmission, user authentication using link keys, and encryption. Bluetooth also offers service discovery, serial port emulation, and Internet protocol (IP) connectivity.

The study of BTnodes has been extended by Siegemund and Krauer (2004), to study integrating handhelds into environments of cooperating smart everyday objects. In their article, the authors show how smart objects can obtain access to resources by spontaneously exploiting the capabilities of nearby mobile user devices. In their application, handhelds join a distributed data structure shared by cooperating smart objects, which makes the location where data are stored transparent for the applications. Smart objects then outsource computations to handhelds and thereby gain access to their resources. As a result, this allows smart items to transfer a graphical user interface (GUI) to a nearby handheld and facilitates the collaborative processing of sensory data because of the more elaborate storage and processing capabilities of mobile user devices.

5.3.2 Efficiency at the Physical and Network Level

Efficiency in wireless sensors, instruments, and networks is important. Conventional instruments and instrumentation systems require two types of wires, one for communication and one for the power supply. Wireless communication eliminates the need for communication wires, thus giving sensors and instruments portability and mobility. Naturally, if sensors and instruments can be made self-powered, the need for wires can be eliminated. This is the reason why the power aspects of wireless systems attract considerable R&D interest.

Many wireless embedded systems, such as multimedia terminals, sensor nodes, and sensor networks, use software constraints on computation needs and communication due to the limited power available in batteries. Total power dissipation is the sum of the power consumption in sensors and transducers, in sensor support circuits, in transceivers and associated circuits, and in microprocessors governed by the intensity of computation.

Advances in digital circuitry, wireless communication, battery technology, and microelectronics are making sensors and instruments smaller, less expensive, and more versatile, reliable, and durable. Although these advancements result in considerable energy savings at the physical level, the systems become much more complex, causing an increase in the energy needed for their operation. Nevertheless, power consumption in many systems can be minimized by energy awareness of the entire system at the individual device level as well as at the network level. Power efficiency and utilization in wireless instruments and instrumentation systems can be viewed in three categories:

- Power efficiency can be obtained by suitable selection of energy-efficient, low-power components and by configuring effective circuits at the individual node level.
- Power efficiency can be obtained by implementing suitable software at the node level, particularly with intelligent sensor systems.
- Power efficiency in the networks can be achieved by effective network management.

The power requirement of each individual instrument depends on its operational characteristics. A new trend in the design of energy-efficient electronic instruments is the management of consumption with the aid of software. The majority of modern instruments employ a digital signal processing, central processing unit, or application-specific integrated circuit (ASIC), and even a communication device, as in distributed sensor networks. In many applications, not all the system elements are required to operate continuously. Therefore a significant power saving can be achieved by adjusting the performance of such circuits according to operational needs.

Power consumption in CMOS logic devices is determined by the following equation (Caldari et al., 2002; Sinha et al., 2002):

$$P = f_{sw} \times C \times V^2, \tag{5.1}$$

where P is power, f is the frequency, and V is the voltage. Lowering the voltage and frequency can significantly reduce the power consumption.

Power efficiency can be provided using a number of methods, including applying power management schemes over the entire network, using low-power mixed-signal circuits (e.g., PIC16LF877 from Microchip Technology), selecting low-power RF transceivers, and optimizing the computational needs of each node or cluster of nodes. Energy use can also be minimized through the use of energy-efficient components.

Dynamic voltage scaling (DVS) is a technique suitable for active power management. In this method, the supply voltage and the frequency of operation are adjusted to the instantaneous needs of the system, resulting in a significant reduction in power consumption (Duarte et al., 2002). The supply voltage is reduced to lowest possible level that meets the required performance criterion at the device level. At the network level, DVS operates in various forms; an example is the central adjustment of the supply voltage and the frequency of operations of each individual device depending on the needs of the entire system.

Dynamic power management is another technique in which the circuit or parts of it are allowed to enter different operating modes such as idle, standby, and stop, and different performance modes (Rong and Pedram, 2002). These modes are entered depending on the instantaneous requirements for system resources.

Energy scalability or energy awareness is also used for power saving. An energy aware system monitors the available energy and dynamically adapts the hardware parameters to meet the latency and performance requirements.

In recent years, a variety of intelligent wireless microsensors have been introduced. These devices are capable of self-maintenance, and robust communication features operate in a maintenance-free manner. Many of these sensors can operate in distributed networks, collaborating with each other or other digital devices. In these devices, the ability to perform low-energy computations is critical, particularly in networks where there are many sensors. The energy used per operation is a key aspect in power consumption digital circuits.

Energy scalability is a new trend that involves the system adapting to time-varying operating conditions. This is not the same as a low-power design approach. The energy aware circuit monitors its available energy and dynamically adapts hardware parameters to meet latency and performance requirements (Zhang and Chakrabarty, 2003).

Today's wireless systems, particularly in distributed sensor networks, use extensive encryption and decryption for secure information flow. The security needs of communicating devices require extensive computation effort and drain a considerable amount of energy (Yuan and Gang, 2002). Energy consumption due to computational requirements can be addressed by various techniques including DVS.

5.3.3 Issues on Communication Protocols and Network Management

In this section, important issues that attract many researchers will be discussed. Since the as hoc networks are relatively new and attract considerable research attention, priority will be given in this area.

Network management has attracted considerable attention since the management of wireless systems is not the same as for conventional networks. With wireless systems, ad hoc networks can be implemented, which is not a trivial task, as it requires carefully developed and implemented algorithms (e.g., Sriram et al., 2004; K. Wang et al., 2004). A mobile ad hoc network is a collection of autonomous mobile nodes that communicate with each other over wireless links. However, since there is no stationary infrastructure, such as a base station, mobile hosts need to operate as routers in order to maintain information about network connectivity (Boukerche, 2004). Thus a number of routing protocols have been proposed and implemented. Some of these protocols are the ad hoc on-demand distance vector (AODV), preemptive ad hoc on-demand distance vector (PAODV), cluster-based routing protocol (CBRP), device status report (DSR), and destination sequencing distance vector (DSDV) routing. In light of these new challenges, researchers are concentrating on a number of issues, some of which will be explained next.

The dynamic and unpredictable environment characteristic of wireless ad hoc networks, coupled with the scarcity of wireless network resources, limits and may even preclude the use of conventional autoconfiguration protocols. The problem of autoconfiguration is exacerbated by the high degree of mobility in ad hoc networks. The dynamic registration and configuration protocol (DRCP), for example, is an autoconfiguration protocol that has been proposed to facilitate dynamic, rapid, and efficient configuration in the unpredictable ad hoc wireless network environment (Vaidyanathan et al., 2003). Such protocols may have problems with performance, convergence time, overhead, and scalability under real-world operating conditions, thus introducing many challenges to researchers.

Intrusion detection is another area of research interest in wireless networks. Wireless networks are known to be vulnerable to intrusion, as they operate in an open medium and use cooperative strategies for network communication. A technique called mobile agent technology efficiently merges audit data from multiple network sensors and analyzes them for the network intrusions. In contrast to many intrusion detection systems designed for wired networks, mobile agent technology implements a bandwidth-conscious framework that targets intrusion at multiple levels. Using this method, the distributed nature of ad hoc wireless network management and decision policies can be taken into account (Kachirski and Guha, 2002).

Quality of service (QoS) in wireless ad hoc networks introduces challenges due to the dynamic operation of these systems. The QoS scheme requires close collaboration between all layers in the protocol stack. At the base of the scheme is the QoS at the media access control (MAC) level. In ad hoc networks, the widely deployed distributed coordination function (DCF) proposed by the IEEE 802.11 standard is a random access scheme, creating a fairness problem. In order to provide a guarantee of minimal throughput and improve the fairness of bandwidth sharing, a group of collision-free MAC scheduling algorithms have been proposed by various researchers. However, when implementing these algorithms, the set of node identifiers that can connect to the network should be known in advance and a neighborhood information sharing scheme is necessary. These algorithms can guarantee minimal throughput and do not rely on per-packet information exchange. Research results show that these algorithms can provide better long-term fairness when compared with IEEE 802.11 (Fang et al., 2004).

Routing of information in ad hoc networks attracts considerable attention due to its complexities. Since the network topology may constantly change and the available bandwidth is very limited in ad hoc networks, calls from the nodes are blocked when a path with the required bandwidth cannot be found in the system. Therefore multiple parallel paths are necessary whose aggregated bandwidth can meet the bandwidth requirement and whose delays are within the required delay bounds. The QoS in the alternative routing paths can reduce the system blocking probability and thus make better use of network resources. The process of searching multiple parallel paths requires at least three parameters—maximum bandwidth, shortest path, and maximum ratio of bandwidth to hops—in order to choose a group of paths whose total bandwidth satisfies the requirement (Wu et al., 2004).

Routing can be managed by distributed routing algorithms or by localized routing algorithms or globalized algorithms (Kuruvila et al., 2004). In distributed algorithms, a node currently holding the packet forwards it to a neighbor that is closer to the destination than itself. This minimizes the ratio of power and/or cost to reach that neighbor as well as the progress (measured as the reduction in distance to the destination) or projection along the line to the destination. In localized algorithms, each node makes routing decisions solely on the basis of the location of itself, its neighbors, and the destination. Globalized algorithms are based on shortest weighted path-based schemes. Much discussion is taking place concerning different types of routing algorithms and their attributes.

Handling of traffic in ad hoc networks requires carefully designed communication protocols. Most routing protocols for ad hoc wireless networks consider the path with the minimum number of hops to be the optimal path to any given destination. However, this strategy may create congested areas, severely degrading the performance of the routing protocols. In some situations, the load over the network may be balanced by selecting a path based on traffic size, while in others methods (Kawadia et al., 2001), mathematical methods are used to generate seeds of pseudorandom numbers to be exchanged between nodes. This exchange of information requires carefully designed software for integration into the OSI reference model.

5.4   Industrial Wireless Sensor and Instrument Networks

The Industrial Technologies Program of the U.S. Department of Energy (DoE) held a workshop in San Francisco, California, in July 2002 to forecast the future of industrial wireless technology. Thirty-four experts representing technology manufacturers, industry end users, universities, national laboratories, and independent consultants concluded that by 2010:

- Wireless sensor networks will become a ubiquitous tool of U.S. industry.
- Sensors and other components of wireless networks will be smart—able to change function in response to dynamic conditions.

- Wireless networks and their components will be autonomous, self-configuring, self-calibrating, self-identifying, and self-reorganizing. They will not require maintenance during their mission life.
- Wireless networks will outperform wired ones in installation and maintenance, upgrading and replacement, failure rate, and rapid commissioning.
- Security in industrial wireless systems will equal or surpass that offered by wired systems due to new encryption technologies and other security measures.
- Wireless systems will be operator independent and will have greater redundancy.
- Wireless sensors will require less power, reducing the cost of operation by 90%.
- Open sensor architecture will allow seamless interoperability at the data level, regardless of the manufacturer.
- The installed cost of wireless systems will be 1/10 as much as in 2002.
- Wireless sensor networks have the potential to revolutionize industrial processing in factory automation, process automation, SCADA, and telemetry systems.
- The use of advanced wireless sensors could improve production efficiency by 10% and reduce industrial emissions by more than 25%.
- Although significant advances in wireless sensor technology are being made, innovations need to be coordinated among developers and the industrial community.
- Wireless sensor systems may run into cultural opposition in some companies, especially when they are capable of autonomous operation.

In the last couple of years we have taken major steps toward reaching these predictions. Vendors offer thousands of products, but end users seem cautious in implementing wireless systems. Nevertheless, today industrial wireless networks exist side by side with wired systems and are gradually replacing them.

Wireless networks in industry are used for long-term monitoring of variables in remote locations, for short-term data gathering on process conditions, to test the economic viability of projects, and for process control, among many others. Wireless networks offer versatility and flexibility in monitoring various applications by providing quick confirmation on process conditions. Wireless field units consist of integrated instruments or sensors, RF communication systems, power supplies, and integrated displays showing the process variable data, history, and network diagnostics such as signal strength, battery life, and communication errors.

In the construction of wireless instrumentation systems, each wireless field unit is assigned an identification code and is capable of communicating with other devices. Although the implementation of ad hoc network topologies is possible, most industrial wireless networks require a base station that communicates with the field devices. This central communicator approach gives added reliability to the operation. The base station exchanges information with the field devices and provides feedback to acknowledge receipt of the information. Usually software runs on a dedicated server to manage the wireless network. Most systems provide a management console for real-time monitoring, configuration management, and reporting.

While setting up a wireless system, there are many factors that need to be taken into account, as illustrated in Figure 5.6a. For example, a project team needs to be set up to look at the level of implementation (e.g., enterprise level, part of the plant, or few devices) and advantages that such a system can offer, such as remote diagnostics and calibration. One important point

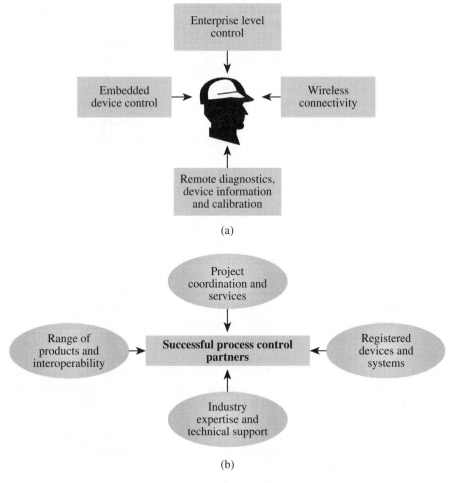

FIGURE 5.6
Important considerations for setting up wireless networks in industry.

is that the workforce needs to be prepared to implement such changes. Once the organization decides to go wireless, other considerations of importance are (see Figure 5.6b) successful project coordination, investigation of the range of available products, the availability of registered devices and even complete systems, and the availability of technical support. Technical support from vendors and other project partners is essential.

Industrial wireless instrumentation networks, just like other networks, operate with OSI reference-based protocols using a variety of band frequencies, including the ISM bands. A complete family of wireless RF sensors, transmitters, loggers, repeaters, receivers, and servers covers frequencies from 418 MHz to 2.4 GHz. Most systems do not require FCC licenses. Multiple frequency transmissions, DSSS, and FHSS techniques are being used successfully in industrial settings. In this section, some examples of industrial wireless implementations are provided.

Wireless sensors and instruments in industry are applied in many areas, including production and control, materials handling, transportation, military applications, security system operations and maintenance, warehouses, etc. Although the application areas are diverse, attention will be paid to two broad types of industrial systems:

- Integration of wireless sensors and instruments with existing fieldbuses and complete wireless networks on the plant floor.
- Wireless built-in tests and condition-based monitoring.

5.4.1   Integration of Wireless Systems to Fieldbuses

The benefits of wireless extensions in industrial networks are well recognized. However, integration of the wireless and fieldbus domains must be capable of retaining real-time operations while maintaining the quality of wired industrial networking solutions. Today, many industries use hybrid transmission media that integrate wired and wireless transmission. Some systems integrate 3G and WLAN radio technology (802.11b, high performance radio local area network [HiperLAN], etc.) with fieldbus technology.

Designing, installing, and using a reliable low-cost industrial wireless data system does not take any more effort than its wired counterpart. Wireless system design must consider the intended range and use of the product so that the system is installed in a location where it will work best. The transmitters and receivers must meet power requirements for each end point for healthy data transmission. Sensor mounting and installation must undergo electrical and mechanical scrutiny for industrial operations.

Large-scale industrial wireless systems are based on four main components: computers, fieldbuses, field devices, and data handlers (e.g., modems, routers, and gateways). The operational dependence of these components on each other is shown in Figure 5.7. In today's industrial wireless networks, modems and gateways play important roles.

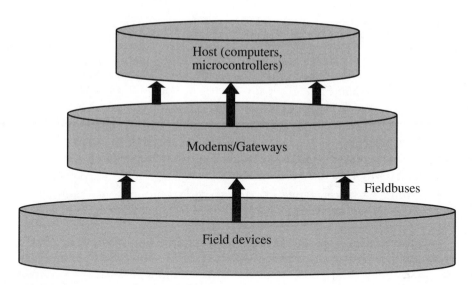

FIGURE 5.7
Structure of a wireless industrial system.

At the moment, modems are the backbone of industrial wireless sensors and instruments, as many industrial devices already contain some communication ports, such as RS-232. Modems are used as part of telemetry for wireless data transfer between field devices and computers and among field devices themselves. The wireless industrial applications of modems can be divided into two main groups: short range and long range. For short-range applications, any of the long-range modems can be used. But for low-cost applications, a variety of short-range modems are available for distances of 100 m or less.

The 2.4 GHz band is accepted worldwide and should be used in applications that require worldwide licensing. There are many industrial devices that are compliant with both Bluetooth and wireless fidelity (Wi-Fi) standards at 2.4 GHz. If the application is in North America, another option is to use 900 MHz products. The 900 MHz band is the most popular in North America and it provides longer distance capabilities. At 2.4 GHz and 900 MHz, many wireless devices are available for the USB, RS-232, RS-422, RS-485, and Ethernet interfaces. However, different countries have different frequency band management and regulations for the use of radio telemetry in industrial operations in both the licensed and license-free radio bands. As guidance, the following are applicable for industrial telemetry systems:

- North America: license-free 900 MHz spread spectrum and licensed 400 MHz to 500 MHz fixed frequency.
- South America: license-free 900 MHz spread spectrum and licensed 400 MHz to 500 MHz fixed frequency.

- Asia: license-free 450 MHz (Singapore, Hong Kong), license-free 220 MHz fixed frequency (China), licensed 400 MHz to 500 MHz fixed frequency (most countries).

- Europe: license-free 433 MHz (all countries), 400 MHz (United Kingdom, Sweden, Finland, Spain, Portugal, Poland, Czech Republic), 869 MHz at 500 mW output 10% duty factor or 5 mW 100% duty factor (in most but not all countries), and licensed 450 MHz (most countries).

- Africa: license-free 433 MHz (some countries) and licensed 400 MHz to 500 MHz fixed frequency.

- Middle East: license-free 900 MHz spread spectrum (some countries) and licensed 400 MHz to 500 MHz fixed frequency (most countries).

Wireless gateways are used in large-scale applications for connecting PLCs, computers, and field devices (Figure 5.8). Gateways are capable of interconnecting different data buses that industrial equipment use. They connect computers, PLCs, digital control systems, and SCADA networks that may be operating with different protocols. Gateways are extensively applied in industrial fieldbuses such as Ethernet, Modbus Plus, Profibus, DeviceNet, DF1, and Modbus. Typical operation frequencies are 220 MHz, 450 MHz, 869 MHz, and 900 MHz.

Ethernet data acquisition systems (EDAS) are being used for many industrial applications. They are based on open architecture standards, allowing for their development and deployment into existing and future plant-wide data capture systems. EDAS units allow users to remotely monitor and control a wide range of sensors, instruments, machines, and processes over

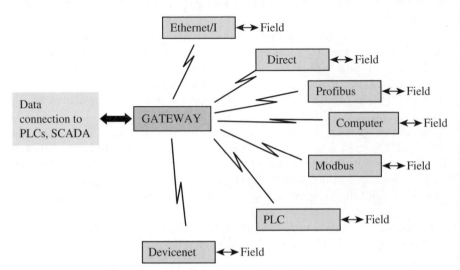

FIGURE 5.8
An industrial network using wireless gateways.

the Ethernet network. EDAS units are used in a range of applications, including machine monitoring and control, networked or stand-alone data logging, environmental monitoring and control, remote data acquisition, and security and access control.

Manufacturers produce wireless transceivers and associated equipment, including extended range WLAN systems and wireless Internet solutions for service providers. In a typical assembly, wireless networking can provide a data rate of 11 Mbs over the air to connect remote stations as far as 30 km away. Multiple security mechanisms work to prevent unauthorized access to both wired and wireless network resources. These devices support point-to-point or point-to-multipoint bridging operations. They also support roaming and a wide range of third-party open air devices. They use FHSS modulation. They also offer wireless bridging solutions for connecting Ethernet LANs between two or more stations.

Most industrial wireless networks currently consist of physically separate devices that require configuration by the end user. Among the many vendors, modern operational industrial wireless networks are exemplified by the products of Elpro Technologies (http://www.elprotech.com). This company offers products suitable for the oil and gas industry, mining and mineral processing, petrochemical and chemical industry, power plants, paper mills, food industries and canneries, manufacturing industries, and so on. The company's product line includes wireless I/O devices, modems, RS-232 and RS-485, Ethernet modems, Profibus, gateways, PLC interfaces, and wireless protocol converters. Their wireless systems are available in licensed and license-free models, 900 MHz FHSS, and 450 MHz, 220 MHz, and 869 MHz fixed frequency.

Plant-wide wireless networking is achieved by using a wireless instrumentation backbone (WIB), as illustrated in Figure 5.9. The WIB comprises wireless J-boxes, each capable of connecting to individual instrumentation signals and/or fieldbuses. Each J-box can exchange instrumentation and control (I&C) signals with any other J-box. The system contains wireless modules or wireless gateway modules. These modules have wireless peer-to-peer communication. Wireless connections or links are configured on a signal-to-signal basis. Any signal at any J-Box can be sent to any other J-Box, in the same way as in an interconnected marshalling box network. Signals are re-created via fieldbus, Ethernet, or Internet. The radio range can be extended by using J-boxes as repeaters. Messages can be routed through multiple J-box units so nodes can talk to one another.

Industrial autonomous networks are another typical example in wireless systems. These systems are still largely in the research phase. They consist of a collection of processes with many autonomous sensors, as illustrated in Figure 5.10. In a system, wireless sensors and instruments are grouped and networked in accordance with their location and proximity to each other. Groups of sensors and instruments are connected to a sensor bus that communicates with some ongoing and independent processes. Each process is

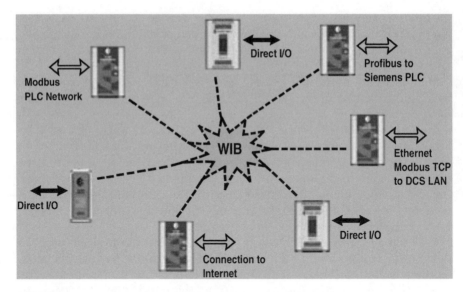

FIGURE 5.9
A typical industrial wireless system. (Courtesy of ELPRO Technologies: http://www.elpro-tech.com/)

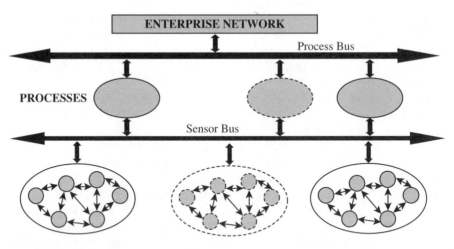

FIGURE 5.10
A typical industrial autonomous network.

then monitored and controlled by the organization's network. There are many other variations of these types of industrial networks.

The Honeywell XYR 5000 wireless system family includes devices for monitoring gauge pressure, absolute pressure, temperature, and ultrasonic noise for detecting steam and gas leaks (http://www.honeywell.com/). The heart of the system is the XYR 5000 transmitter. This wireless transmitter operates with ModBus and uses the host/master principle for networking.

It uses 11-bit characters per frame: 8 data bits, even parity, and 1 stop bit. The XYR 5000 can communicate with a variety of wireless devices on 4 mA to 20 mA or RS-232 or RS-485 outputs. It uses FHSS and the distance varies depending on the speed of the data transmission: 150 m for 76.8 kbaud, 300 m for 19.2 kbaud, and 650 m for 4.8 kbaud. The whole system is supported by Wireless Management Tool Kit software.

The XYR 5000 line includes an analog input interface for adding wireless capabilities to 4 mA to 20 mA devices. The instruments transmit measurements to a base station connected to a control system or data acquisition device such as a recorder or PC. Each base station accepts signals from up to 50 transmitters. The base station provides Modbus or 4 mA to 20 mA analog signal output. Honeywell largely offers Foundation Fieldbus products for measuring pressure, temperature, level, and flow, control actuator valve positioners, and complete control systems.

Honeywell XYR 5000 systems consist of smart field devices such as a wireless pressure transmitter. This pressure transmitter operates in the 902 MHz to 928 MHz license-free ISM band and is powered by C-size 3.6 V lithium batteries with a life expectancy of five years.

Other vendors offer industrial wireless sensors and instruments. For example, the OM-CP-RFTC4000A, from Omega Engineering Inc. (http://www.omega.com/), is a miniature, wireless, battery-powered, stand-alone, thermocouple-based temperature transmitter. This compact and portable device measures and transmits temperature data. When enabled, the wireless transmitter sends readings back to a host computer where the data can be analyzed in real time. Readings are also logged to the device's memory for added data security. A slide switch allows the transmitter to be turned on and off without affecting the operation of the device. Data are received at the PC using the OM-CP-RFC101A receiver, which attaches directly to the serial port. This device has the following characteristics: transmission rate 4800 baud, RF carrier frequency 418 MHz, range 36 m, and it is supported by Windows 95/98/ME/NT/2000/XP. It is powered by a 3.6 V lithium battery with an expected battery life of one year at taking readings every minute at about 25°C.

Wireless meter readers are unlicensed transponders for gas, water, and electric meter reading applications. The transponder is a low-power device that operates as part of a remote meter reading system that monitors utility meters. Wireless meter readers represent fully automated industrial wireless technology. In a typical application, the transponder collects usage data and stores it in preparation for interrogation. Usage information is then retrieved with an RF communication link.

In a product offered by Innovative Wireless Technologies (http://www.iwtwireless.com/), the wireless meter reader has a real-time clock generated by the microcontroller. It has the ability to store the last 30 days of utility usage. A low quiescent current microcontroller and regulator offers superior sleep mode current consumption that translates into a battery life of 16 years. Some wireless meter readers have a range of more than 2000 m.

Cellular networks are another way of networking wireless sensors and instruments and are also used for wireless meter readers. For example, Meter-Master (http://www.meter-master.com/ms/) has a variety of wireless meter readers that are capable of sending text messages (SMS) using GSM networks. A device called the Cello is a cross between a cell phone and a submersible meter interface unit. It can operate in poor coverage areas such as underground vaults and basements.

LonWorks fieldbuses, from Echeleon Corporation (http://www.echelon.com/), integrates wireless devices with a fieldbus using Modbus/TCP universal gateways. An IEEE 802.11b-based wireless gateway can poll a Modbus slave IP address, which can be held by any number of LonWorks devices. Gateways operate as clients or servers with master and slave functionality on an object-by-object basis. Wireless gateways find applications in building automation, light monitoring, remote alarm monitoring, utilities, backup power facilities, semiconductor fabrication, lighting control, telecommunication, light industrial operations, etc.

Echelon Corporation also offers the RRX04F wireless switch for use in LonWorks networks. They are available with eight output channels; each can be programmed for toggle or momentary action and can issue ON/OFF, PANIC, and SCENE outputs. The RRX04F receiver can be mounted horizontally or vertically with surface or in-wall mounting. The system incorporates a code hopping decoder for secure remote keyless entry (RKE). It utilizes a patented code hopping system for high security. The system comes complete with a two-channel rolling code transmitter. An integrated printed circuit board at the rear of the device incorporates a neuron chip and a transceiver.

5.4.2 Wireless Built-in Tests and Condition–Based Maintenance

Built-in tests (BITs) can be described as a set of evaluation and diagnostic tests that use resources that are an integral part of the system under test. BITs can be applied at the circuit level, module/assembly level, or system level. Circuit-level BITs use analog and digital methods. These tests are characterized by well-developed standards and technologies (IEEE 1149) and signature analysis and tools. Module-level BITs are supported by programmable instrumentation. System-level BITs are self-contained and are supported by sophisticated hardware and software. At the system level, BITs may require synchronized data from many signals, and parameters and thresholds may have very short temporal dependence requiring immediate attention.

Radio frequency data communication is extensively applied in BITs in which the sensors are embedded in operational systems, such as grinding wheels, rotating machinery, industrial systems, or even implanted in living organisms. The information gathered from the sensors is transmitted by a built-in RF transmitter to a nearby receiver. The quality of BITs is improving steadily because of the availability of intelligent and other sophisticated sensors, instrument software, interoperability, and self-test effectiveness.

An example of a wireless device that is suitable for BITs is configured by Systran Federal Corp. (SFC) (http://www.systranfederal.com). This unit contains a lightweight, handheld unit called PALM-IT that is used for performing flight-line diagnostics and maintenance operations. The heart of the system consists of a five-function PC card supporting the following interfaces: Mil-Std-1553, IEEE 488, 10Base-T Ethernet, and Fortezza encrypted wireless networking. The first two interfaces are used to interface with avionics on the aircraft, the others are used to communicate with ground-based equipment. PALM-IT was designed as a complete hardware and software solution for flight-line operations. It provides the tools needed for the sponsor to develop custom applications for uploading and downloading operational flight programs (OFPs), initiating BITs on the avionics, transferring diagnostic information between the aircraft and ground-based equipment, and performing other maintenance operations.

Another example is condition-based maintenance (CBM), which is a program maintained by many organizations. CBM is of particular importance to the Navy, Air Force, Army, space exploration, NASA, oil and mineral exploration, transportation systems, railway systems, and other types of industry. In CBM, data are collected on vibration, temperature, speed, and other variables of operational devices and machines. By the use of appropriate analysis software, the health of the machine can be assessed and predictions made on the possibility of machine failure, its lifetime, servicing and maintenance requirements, and so on. CBM helps prevent unplanned downtime for a machine, resulting in optimum production.

A version of CBM is the use of on-line surveillance systems, which provide current information on a machine's condition. The on-line system includes hard-wired or wireless sensors connected through multiplexers, which are networked to a database computer. Depending on the size of the operation, a full implementation of an on-line surveillance system may be very complex and costly. It may need a great deal of planning and may be difficult to install. Apart from the up-front data collection hardware, software, and training costs, labor remains a continuing expense.

A critical question is how often does equipment need monitoring? With well-designed and managed programs, data collection by CBM may be made cost effective. On-line surveillance systems are more cost effective than data collectors when the machinery is critical to a process, has a history of failure, or operates beyond its original design. In this case, frequent condition monitoring can change the way maintenance and production decisions are made.

Wireless CBM data gathering systems are viable because of advances in wireless technology and the miniaturization of sensors and monitoring circuitry. Since machine health monitoring has serious implications for production continuity, the widespread implementation of wireless condition monitoring systems may be justified. With up-to-date machine conditioning, maintenance, and production, decisions can be made to optimize production, and manufacturing facilities can operate more efficiently.

Elimination of cables in wireless systems is a natural progression for CBM and adds a degree of flexibility. Several companies have developed wireless CBM systems, such as Wilcoxon's BlueLynx and Oceana Sensor's ICHM 20/ 20. In addition, Rockwell has the HiDRA, and SKF promotes its wireless systems. These systems connect hard-wired sensors to a radio hub, but still require significant cabling.

For example, Oceana Sensor (http://www.oceanasensor.com/) offers a range of wireless products that can be used in condition-based monitoring systems. The company offers different network products such as the SHM-INDUSTRIAL network, ICHM 20/20 network, SHM PDA network, and central workstations that can be used for CBM. The ICHM family of products is built on Bluetooth wireless technology and is suitable for many other wireless industrial network applications. The ICHM 20/20 accommodates various sensor types through in-line signal conditioning interface modules (SCIMs). The on-board digital signal processor on the SCIM can handle signal processing, providing feature extractions of the raw data prior to wireless transmission. It utilizes 5 V DC line power to power the device and charge an optional internal lithium ion battery for backup. Power can be supplied by either 110 V AC-to-5 V DC converters or a DC-to-DC converter. This company offers various sensors for diagnostic purposes, such as accelerometers, temperature sensors, and acoustics sensors.

Plug-and-play wireless CBM systems provide ad hoc networking capabilities in conjunction with appropriate wireless sensor nodes. Ad hoc networks allow additional senor nodes to be added. The plug-and-play transceiver receives the information from the wireless sensor network and connects them to a central computer. Most enterprises have some form of existing network based on Ethernet, which provides flexibility when compared to other fieldbuses such as Modbus, DeviceNet, and Fieldbus. With the use of wireless networks integrated with Ethernet for CBM, the data can be transferred and viewed anywhere in the world. This is particularly important for organizations that have plants and operations scattered throughout the world. Once the information on factory conditions is on the network, the information can be viewed in a central location that monitors distributed operations. Within this context, application software must be usable by multiple users.

5.5 Wireless Human Health Monitoring and Environmental Applications

Wireless sensor and instrument networks are extensively applied for monitoring human health. Human health applications include telemonitoring of human physiological data, tracking, and monitoring inside hospitals. Typical functions include glucose level monitoring, organ monitoring, and general health monitoring.

Wireless sensors and instruments are applied in environmental condition too for monitoring, tracking, and health monitoring of wildlife and domestic stocks. Large and dense networks of sensors are deployed to monitor a wide variety of real-world phenomenon and distributed physical processes. Wireless sensor and instrument networks are used in environmental monitoring, agriculture and forestry, seismic monitoring, air quality, and military applications.

5.5.1 Wireless Human Health Systems

Today wireless sensors and instrument networks in human health are applied in a variety of forms ranging from implantable wireless medical devices to remote health monitoring systems. Since they need to be small in size, the integration of communication hardware and software in these devices presents many challenges in protocol implementation, application development, and security issues.

Wireless implantable biomedical sensors have the potential to revolutionize medicine. Smart sensors have been developed for several biomedical applications, including glucose monitoring, microlevel organ activity monitoring, and retina prosthesis. These devices have the ability to communicate with an external communication system or a base station via a wireless interface. The limited power and computational capabilities of smart sensor-based biological implants present research challenges. These sensors must be biocompatible, fault tolerant, energy efficient, and scalable. Also, embedding the sensors in humans adds additional requirements. They must be ultrasafe, reliable, work trouble free in any geographic location, shouldn't restrict the movements of their human host, and should need minimal maintenance. These requirements necessitate application-specific solutions that are vastly different from traditional solutions.

Recent progress in biomedical sensing technologies has resulted in the development of several novel sensor products and new applications. Modern biomedical sensors developed with advanced microfabrication and signal processing techniques are becoming inexpensive, accurate, and reliable. A broad range of sensing mechanisms has significantly increased the number of possible target measurands that can be detected. The miniaturization of classical measurement techniques has led to the development of complex analytical systems, including such sensors as the biochemlab-on-a-chip. This rapid progress in miniaturization will likely have a great impact on the practice of medicine as well as advances in biomedical research. Currently electrochemical, optical, and acoustic wave sensing technologies have emerged as the most promising biomedical sensor technologies.

Implanted wireless sensors are a good example of modern biomedical sensors. One such sensor is the artificial retina created for visually impaired

268 Wireless Sensors and Instruments

people. In this system, a Smart Sensors and Integrated Microsystems (SSIM) project, a retina prosthesis chip consisting of 100 microsensors is built and implanted in the human eye. This allows patients with no vision or limited vision to see at an acceptable level. Wireless communication is needed for feedback control, image identification, and validation. The sensors produce electrical signals that are converted into a chemical response, mimicking normal retina operation. The device uses wireless communication from a camera embedded in a pair of eyeglasses. The challenges in this application are low power in the sensors and transmitter, limited computation capability, material constraints, lifetime, robustness, scalability, security, and regulatory requirements. The communication pattern is deterministic and periodic, and TDMA works best in this application for energy conservation (Schwiebert et al., 2002).

Another example of implanted wireless sensors is an underskin sensory system that provides real-time information on the position of the lower extremities and the trunk of a patient. A typical sensor pack is microprocessor controlled and contains a miniature rate gyroscope and a pair of two-dimensional accelerometers. A navigator, which is worn on the body, acts as the controller. The navigator runs software customized for each patient. The navigator collects body position data from the sensors, allowing it to implement closed-loop control strategies. Menu-driven operation is displayed on a color touch-screen display. The communication takes place via a wireless RS-232 protocol. The navigator powers and controls the implanted sensors via a transmit coil that is held in place on the skin surface, thus forming an RF data link.

Mobile health monitoring is another application of wireless sensors and instruments. Mobile health monitoring is useful in emergency medical care and disaster response, and allows patients to remain mobile during monitoring. In emergency medical care and disaster response, wireless sensors are used to monitor the vital signs of patients, such as heart rate and oxygen saturation. These telemetric systems are designed to measure physiological parameters in an unobtrusive way and transmit the data for remote monitoring. This provides ambulatory patients with freedom of movement while their health is being monitored. Such systems are useful for monitoring the vital signs of service personnel (e.g., firefighters working in hazardous environments). Wireless sensors can also be used for tracking patients and rescue workers. There still are many challenges that must be addressed in developing appropriate sensors and instruments, common protocols, and software frameworks to integrate a range of devices to operate uniformly and promptly.

Mobile health monitoring in Sweden (http://www.thelocal.se/) is an example of a wireless health system. A service, called BodyKom, has been developed by TeliaSonera Sweden in association with Hewlett Packard and the Swedish software company Kiwok. BodyKom functions by means of a communication device that is connected wirelessly to a number of sensors on a patient's body. The device communicates with health care personnel over TeliaSonera's nationwide mobile network, constantly monitoring the patient's health. In this system, hospital and health care staff are able to remotely and continuously monitor patients that do not require in-patient treatment. If the sensors detect a change in the patient's health, or if the patient notices a change, the system can automatically dispatch an alarm to an on-call doctor or to members of the patient's family. BodyKom is currently being used to monitor the pulse of patients. In the future the service will also be offered to patients with diabetes, asthma, and other illnesses that might require emergency assistance.

Medtronic, Inc. (http://www.medtronic.com/) has a system called the CareLink Network. This network allows doctors to monitor patients by taking data from pacemakers and other implanted devices. These devices can be used to sense fluid buildup in the lungs, electrocardiogram (ECG) changes, and other physiological data. The collected information is then transmitted over a phone line to a secure Internet site accessible only by doctors. Pacemakers can also act as implanted cardioverter defibrillators (ICDs) to stabilize an irregular heartbeat.

Mobile cardiac units are an example of a wireless health monitoring system. CardioNet (http://www.cardionet.com) has developed a mobile cardiac outpatient telemetry system called CardioNet. The CardioNet system consists of a small three-lead ECG monitor worn as a pendant around the neck or on a belt clip, and a PDA-like device. The ECG monitor sends its data via a 900 MHz wireless link to the PDA, which evaluates and stores the information. Information related to any irregularities in the heartbeat of the patient can be forwarded to a central station via a cellular or metropolitan area network for further action by health care personnel. This information can then be used for diagnosis, prevention, and treatment.

An ICD with a mobile remote data transmitter is being manufactured by Biotronik (http://www.biotronik.com). The ICD automatically transmits diagnostic data to the attending physician via the Biotronik Home Monitoring System. The data are transmitted through a cell phone-like device over a GSM network to a service center. At the center, the data are analyzed and the physician is notified immediately in case of an emergency. The system uses a frequency band of 402 MHz to 405 MHz, which passes through the skin without significant attenuation.

Human biotelemetry helps those working in hazardous areas. Instruments equipped with time-frequency (TF) transceivers are used to detect the potential onset of distress, prompting precautions or emergency action to be taken. Human biotelemetry is applied in manned space flights to monitor the vital signs of astronauts. During the first manned landing on the moon, the voices of the astronauts were combined with 900 other signals, some of which were the physiological parameters from the astronauts.

An example of human biotelemetry is the telemetric nurse, based on biomedical sensors and instruments. In these systems, sensors and instruments monitor the patient's vital signs. The data generated are transmitted by transceivers to a base station for observation and recording. The base station receives the data and analyzes the information and displays.

Cardiac pacemakers rely on telemetry to improve the quality of life of patients who wear them. The telemetry process is initiated by the cardiologist by sending a special signal to the pacemaker; normally a transducer is placed on the patient's chest. By using transceivers, the physician is able to remotely program different codes or pacing algorithms into the implanted pacemaker. Typical data that pacemakers normally generates include voltage, current, width, charge, and energy of the delivered pulses; the count of sensed and paced events and an ECG; the impedance, current, and voltage levels of the batteries; and the impedance of the lead contacts. All the performance parameters can be programmed to be read back or interrogated, allowing the physician to confirm the programming of the pacemaker.

Wireless home care systems for the elderly are attracting considerable R&D attention. It is anticipated that the worldwide population of those over 65 years of age will reach 761 million by 2025, more than double the population in 1990. Companies such as General Electric, Hewlett Packard, Honeywell, and Intel (Dishman, 2004) teamed up with the Center for Aging Services Technologies (CAST) (http://www.agingtech.org) in 2002 (Ross, 2004). CAST was formed to encourage collaborative aging-related technology. The idea is to deduce the actions of older people in their homes through a network of wireless sensors and use the information to help people comply with doctor's orders, enable remote care giving by family and friends, and detect early signs of disease.

The key technology in a wireless home care system is a small battery-powered sensor, called a mote. These sensors, developed at the University of California, Berkeley, and Crossbow Technology, Inc., organize themselves into wireless networks, sharing data with one another and with other computers. The software in the computers is able to analyze the information coming from the motes, record the routine tasks, and act on information if an alarming situation occurs.

5.5.2 Wireless Environmental and Habitat Monitoring Systems

Habitat monitoring is an important application of wireless sensor networks. In these applications, low-power wireless sensor networks are employed for long-term information gathering on various subjects. R&D efforts in this area concentrate on issues such as the network behavior regarding hardware and software, node failures, radio transmission capabilities, interference, shielding of components from various environmental elements such as rain and fog, network topologies, self-organization capabilities, power requirements, and energy efficiency.

The University of California, Berkeley Mica mote study was begun in 2002 to investigate the operational characteristics of wireless habitat monitoring systems. In this study, 32 motes were placed in the area of interest. These motes were grouped into sensor patches that transmit sensor readings to a gateway (CerfCube) that is responsible for forwarding data from the sensor patch to a remote base station through a local transmit network. The base station replicates the data every 15 minutes to a satellite-based database, for information logging. The sensor network can be accessed locally for adjustment and monitoring of the network components or remotely via the satellite link.

In this application, the Mica mote is the heart of the system, which uses an Atmel ATmega 103 microcontroller running at a speed of 4 MHz. A 916 MHz RF transceiver provides bidirectional communication at 40 kbps. The system runs on a pair of AA-size batteries. A Mica Weather Board is stacked on the processor board using a 51-pin extension connector and contains temperature and humidity sensors, and a photoresistor, barometer, and thermopile.

PODS is a Remote Ecological Micro-Sensor Network is a research project at the University of Hawaii that built a wireless network of environmental sensors to investigate endangered plants and wildlife. This system consists of a computer, radio transceiver, environmental sensors, and digital cameras. System information is relayed to an Internet link. Bluetooth and 802.11b are used for MAC and data are delivered to IP packets. Energy efficiency is identified as one of the design goals, and ad hoc routing protocols called multipath on-demand routing (MOR) were developed. The user can access the system from the Internet to observe weather data that are updated every 10 minutes and image data collected once every hour.

In this study, sensor placement strategy was investigated. The sampling distance and communication radius were identified as key parameters. The topologies consisted of triangle, square, hexagonal, ring, star, and m-tiles. The sensor placement strategy is based on three goals: resilience to a single point of failure, area of interest to be covered by at least one sensor, and minimum number of nodes.

Wildlife biotelemetry makes otherwise impossible experiments feasible for tracking and monitoring wildlife. The application of wildlife telemetry is particularly important in those situations where it is desirable to leave the subjects in a relatively normal physiological and psychological state by not interfering with their normal activities. The fact that a very small radio transmitter can be swallowed opens up vast possibilities for areas of study. Also, the absence of electrical leads connecting the sensor to the subject's body is especially important in long-term studies, as they can greatly interfere with the subject's normal activities.

Implanted transmitters have been extensively used to study the habits and patterns of wildlife. In these types of applications, the life of the battery and the stability of the recording transducer are critical to the success of the project. This type of approach has been used to study peristalsis in cold-blooded wildlife by using a simple transmitter that measures pressure and temperature that is implanted in a mouse that is subsequently swallowed by a snake. The researchers recorded temperature and pressure data for many days until the transmitter ceased operation.

The range of the transmitter is another important parameter to consider when using telemetry to study wildlife. The distances can range from a few centimeters in laboratory experiments to several miles when tracking wildlife. Another critical parameter is selecting the correct transducer; the transducer must accurately sense the desired range of physiological changes while also being resistant to corrosion from body fluids. A third important parameter to consider is the temperature range the transmitter must withstand. Transmitter battery life can be greatly affected by the operating temperature.

DolphinEAR 500 (DE500) is an example of wireless hardware and software used in habitat and wildlife monitoring. This device is offered by a company called DolphinEAR (http://www.dolphinear.com/). The DE500 is a combination of a hydrophone, preamplifier, and mobile telephone interface. It plugs into the accessory connector of any handheld mobile telephone and allows the user to dial up and listen to the hydrophone from any telephone in the world. This audio stream can be connected directly to a computer for storage on a hard disk or it can be recorded in a conventional manner on minidisc or standard tape equipment. Any standard low-cost digital phone (such as GSM or CDMA) or older style analog phone can be used. It does not require a special data connection. The user simply dials the telephone number of the hydrophone to initiate a monitoring session. Each device is supported by software that produces spectrograms. The software allows the user to display underwater sounds in real time.

The NanoTag series of microprocessor coded tags are an example of the sensors used for tracking wildlife and livestock. Lotek Wireless (http://www.lotek.com/) offers a wide range of products for habitat monitoring that are suitable for radio, acoustic, archival, and satellite monitoring applications in terrestrial, freshwater, marine, and avian habitats. NanoTags bring three distinct technologies together: radio, digital processing, and infrared systems on a 2.8 mm^2 ASIC. The radio transmitters are used with digitally encoded systems based on a proprietary coding scheme. This coded telemetry system allows up to 212 transmitters to be assigned on a single frequency, while retaining the ability to identify the operation of individual transmitters. Coded sensor transmitters include temperature, pressure, and motion sensing. These sensors are available either individually or in combination on transmitter platforms of radio, acoustic, and combined acoustic/radio.

5.5.3 Environmental Observation and Forecasting Systems

Environmental observation and forecasting systems (EOFS) are distributed systems that cover large geographic areas. These systems monitor, model, and forecast physical processes such as environmental pollution, weather, volcanic activity, flooding, tides, earthquakes, tsunamis, etc. EOFS consist of three main components: the sensor stations, the distributed network, and a centralized processing station. The central processing station is normally a computationally intensive place since, in many applications, the volume of data gathered from the sensors is very large. There are many examples of operational EOFS ranging from small-scale systems to global observation schemes such as the Global Earth Observation System of Systems (GEOSS) and the Mediterranean Forecasting System Toward Environmental Predictions (MFSTEP).

The MK III is a typical example of a small-scale wireless outdoor weather station, available from American Weather Enterprises (http://www.americanweather.com/). The MK III is solar powered, with a transmission range of up to 140 m using a frequency of 315 MHz. The sensor assembly includes rain gauges, temperature sensors, wind speed and direction sensors, and humidity sensors. The data collected by individual sensors are transmitted to one or more receivers that include computer software for analysis and display.

The Automated Local Evaluation in Real Time (ALERT) system is also a typical example of an environmental observation and forecasting system. It was built in 1970s and is probably the most well-known wireless sensor network. ALERT sensor sites are equipped with meteorologic/hydrologic sensors that collect information on water level, temperature, and wind. The data are transmitted by radio links to various base stations. Currently ALERT is used across most parts of the western United States, mainly for flood warning.

CORIE is a prototype EOFS for the Columbia River. In this system, 13 sensor nodes are deployed across the river. The sensor stations are fixed on a pier or a buoy. One station drifts offshore and power for this station is provided by solar energy. Sensor data are transmitted to an onshore base station via a wireless link.

The Geostationary Operational Environmental Satellite R series (GOES-R) program is a typical environmental observation and forecasting system that will incorporate new instruments into the next series of satellites with an initial launch readiness of 2012. The GOES-R system will perform a range of missions, including storm and flood warnings, tropical cyclone warnings, hydrologic forecasts, water resource management, ocean surface and internal structures forecasts, solar and space environmental forecasts, domestic and military aviation forecasts, ice condition forecasts, seasonal and interannual climate forecasts, environmental air quality monitoring and emergency response, fire and volcanic eruption detection and analysis, and long-term global environmental change assessments.

5.6 Radio Frequency Identification

This book would not be complete without a discussion of RFID since this is becoming an important application of wireless sensor and instrument networks. RFID is a generic term for a bunch of technologies that use radio waves to identify people and objects. In its basic structure, RFID is a tiny microchip attached to an antenna. This assembly is called an RFID transponder or an RFID tag. The antenna enables the chip to transmit identification information to an electronic receiver or reader. The receiver converts the radio waves from the RFID tag into digital information that can be processed and put into a useful form. Depending on the type of microprocessor used, RFID tags have the potential to contain much more information than just the identification number about a person or object.

Radio frequency identification systems have been around since the 1970s. A complete RFID system consists of an RFID tag (passive, active, semipassive), an antenna or coil that can read/write data to and from the tag, a tag reader, tag programming stations, and sorting equipment. Usually the antenna and reader are integrated into one product. Advanced large-scale integration chips are the key component. RFIDs are categorized by their RF: low-frequency tags (125 kHz to 134 kHz), high-frequency tags (13.56 MHz), ultrahigh frequency (UHF) tags (868 MHz to 956 MHz), and microwave tags (2.45 GHz and 5.8 GHz).

An active RFID tag uses an internal power source within the tag to continuously power the tag and its RF communication circuitry. This allows extremely low-level RF signals to be received since the reader does not power the tag. Active RFID tags are continuously powered and are normally used when longer read distances are required.

Passive RFID relies on RF energy transferred from the reader to power the tag. They reflect energy from the reader to receive and temporarily store a small amount of energy from the reader signal in order to generate a response. Passive RFID requires strong RF signals from the reader and the RF signal strength returned from the tag is constrained to very low levels by the limited energy. They are best used when the tag and interrogator are close to one another.

Semipassive RFID uses an internal power source to monitor environmental conditions, but requires RF energy transferred from the reader, similar to passive tags, to power a response. Semipassive RFID tags use a process similar to that of passive tags to generate a response. They differ from passive tags in that semipassive tags possess an internal power source that allows the tag to complete other functions, such as monitoring of environmental conditions (e.g., temperature, vibration, shock, etc.), that extend its operation capabilities.

Radio frequency identification still has a number of problems to be solved before their full-scale application. One problem is in the reading of tags.

Passive and semipassive tags are largely based on the technique of backscattered modulation to communicate with readers. Backscattered signals are much weaker than transmitted signals. Therefore the presence of a strong transmitted signal from the reader creates a blind spot and reduces sensitivity, thus making it hard to identify tag signals. Researchers have offered a number of solutions, such as the use of multichannel receivers backed up with appropriate filters.

Another problem is the security of information stored on the tags. Since the information on a tag is simply drawn out by a reader, an RFID system may infringe on consumer privacy. Although identification of an RFID tag can be encrypted, it is possible to trace an object, even if its contents are not understood, by pursuing specific information. Researchers are concentrating on security and privacy protection that prevents pursuit of a tag using various methods such as reencrypting tags using universal reencryption techniques. Security can be handled in a number of ways. Security gates can be used to determine a tags security status or the tag can contain security codes that can be interpreted by reader stations.

Standardization of RFID products is an ongoing process. There is no global public body that governs the frequencies used for RFID. In principle, every country can set their own rules, but the main bodies governing frequency allocation for RFID are the FCC in the United States, the Department of Communication in Canada, the European Telecommunications Standards Institute (ETSI) in Europe, the Ministry of Public Management, Home Affairs, Posts and Telecommunications in Japan, and the Ministry of Information Industry in China. An organization called EPCglobal (http://www.epcglobalinc.org) is working on an international standard for the use of RFID and electronic product code (EPC) identification of any item in the supply chain for companies in any industry, anywhere in the world.

The U.S. Department of Defense (DoD) recognizes the potential applications of RFID and its implications in security matters and has issued RFID regulations. In February 2003 DoD partnered with industry and held strategy sessions with key industry associations and standards experts to shape a policy on RFID. Following the initial meeting, RFID summits were held in December 2003 and April 2004 to introduce important concepts of a pending DoD policy that would become effective in January 2005. Today DoD uses a frequency range for the tags of 860 MHz to 960 MHz with a minimum read range of 3 m. The classifications of the accepted tags are class 0 64-bit read only, class 1 64-bit read-write, class 0 96-bit read only, and class 1 96-bit read-write.

The application areas of RFID are diverse. Low-frequency RFID tags are commonly used for wildlife and domestic livestock identification and tracking and antitheft systems in automobiles. High-frequency RFID tags are used in library books and bookstores, building access control, airline baggage tracking, identification badges, and so on. UHF tags are used commercially in pallet and container tracking, transport vehicles, and truck tracking.

Microwave tags are gaining popularity and are being used in long-range identification and tracking applications.

Driven by aggressive initiatives launched by retail and government organizations in the United States and Europe, RFID is gaining acceptance for use in supply chains. In 2004 Wal-Mart announced that the retailer's top 100 suppliers would be delivering cases and pallets with RFID tags by January 2005. This initiative kick-started—and in some ways redirected—progress on supply chain technology that had been of interest to researchers at a handful of technology vendors, academic labs, and large businesses. Companies such as Hewlett-Packard and SAP are introducing products and services to accelerate RFID standards, and tagged boxes are shipping in real-world trials.

The introduction of RFID in the supply chain is widely regarded as a replacement for barcodes in tracking goods. Unlike barcode systems, it doesn't require line of sight or any other orientation to read a tag. This reduces the manual component required for the success of the system. RFID tags also possess a memory component that enables them to store information about the features of a product and its manufacturing details. This information can be embedded onto the tag at the time the item is manufactured. Thus RFID holds the promise of revolutionizing global supply chain operations and the way many companies do business.

In industry, RFID also finds applications on the plant floor for asset maintenance. It is used to track raw material production, procurement of components, manufacturing and assembly, purchases, deliveries, use, maintenance, repair, and disposal or recycling. Many large companies that have invested heavily in enterprise asset management (EAM) and want to get more out of their investment are pairing RFID with mobile software, allowing work orders to be updated with key data points that are transferred to the tag. This information is routed back to the EAM to make reporting more meaningful, maintenance more effective, and reliability planning more efficient.

A typical RFID tag is illustrated in Figure 5.11. This particular product, part no. RF-HDT-DVBB-N0, by Texas Instruments (http://www.ti.com), is a 13.56 MHz encapsulated transponder. It is used for applications that require a rugged transponder able to withstand harsh environments with temperatures from –25°C to 220°C. The encapsulated transponder is compliant with the ISO/IEC 15693 standard, a global open standard that allows interoperability of products from multiple manufacturers.

A series of typical RFID transponders and tags are illustrated in Figure 5.11. All these transponders are compliant with the ISO/IEC 15693 standard. The rectangular transponder in Figure 5.11a and the circular transponder in Figure 5.11b are called inlays. They operate at 13.36 MHz and are suitable for use under different regulatory and noise environments. They contain 2 kbits of user memory organized in 64 blocks and are used in variety of applications including product authentication, ticketing, library applications, supply chain management, CD/DVD labeling, and so on. Since they are thin and flexible, they can easily be converted into paper or plastic labels.

| a) Circular inlay | b) Mini inlays | c) Circular transponders |

FIGURE 5.11
An RFID tag.

The RFID transponder in Figure 5.11.c contains a 24 mm circular inlay and is housed within a PVC substrate. It is designed for use in harsh environments with temperatures ranging from 25°C to 220°C.

Another example of an RFID product is the VeriChip, by Applied Digital (http://www.adsx.com). The VeriChip is a miniaturized, 12 mm × 2.1 mm, implantable RFID device. The chip contains a unique verification number, which is captured by briefly passing a proprietary scanner over it. A small amount of RF energy passes from the scanner, energizing the transponder, which then emits an RF signal transmitting the verification number. The company is promoting VeriChip in a variety of applications in defense, homeland security, and other secure-access applications. Potential areas of use include controlling authorized access to government installations and private-sector buildings, nuclear power plants, national research laboratories, correctional facilities, sensitive transportation resources, etc. The company recently unveiled VeriPass and VeriTag, which will allow airport and port security personnel to link a VeriChip subscriber to his or her luggage during check-in and on the airplane. The concept of using VeriChip can also be extended to include a range of consumer products such as PCs, laptops, cars, cell phones, and homes.

The ALB-2484 is another example of an RFID tag, by Alien Technology (http://www.alientechnology.com). This company produces a range of UHF and microwave frequency tags. The ALB-2484 is a long-range, battery-powered, backscatter tag that can be interfaced with any external sensor such as temperature sensors, tamper indicators, shock sensors, etc. Since the tag does not have an active transmitter, it is said to be quiet, thus it does not require FCC certification. The tag has many additional features such as analog and digital I/Os, external sensors including temperature and humidity, extended memory, tamper detection, and shock and vibration sensing. Some application examples suggested by the manufacturer are long-range identification, vehicle asset tracking, security access systems, supply chain automation, cold chain management, passive tag data storage for hierarchical asset tracking systems, and so on. The reader for this tag can be interfaced either locally or through a local area network (LAN) interface to remote servers. Input control lines on the reader allow for trigger inputs, which turns on the RF only when goods are present, thus alleviating interference with other RF sources. Output control lines enable the operation of a gate or door when a valid tag is interrogated. Communication takes place via RS-232 at 115,200 baud. The LAN interface is realized by a TCP/IP or alien dynamic assimilation protocol. There are two separate antennas for transmitting and receiving. It has visual indicators for power, RF, transmit, receive, sniff, and lock.

5.7   Consumer Products and Other Applications

The consumer market is playing an important role in the development and deployment of wireless products and networks. For example, cell phones have been one of the major contributors to modern communication systems and mobile network technologies. Bluetooth was first targeted at the consumer market and then found many industrial and other applications. In this section, wireless sensor and instrument networks as applied to consumer products are discussed and a few examples given. Other examples of wireless sensors and instruments applied in unusual areas are also provided.

5.7.1   Wireless Consumer Products

Wireless consumer products include handheld and body-worn computers, wireless credit card readers and receipt printers; home entertainment systems and electronic games; home automation systems; automotive keyless entry systems; wireless security systems; sports, hobby, and recreational equipment; RFID tags and access control badges; and health care and medical wireless devices. Some examples of wireless consumer products are discussed below.

Wireless headphones are a typical example of a wireless consumer product and they are manufactured by many companies. For example, the MDR-RF960RK is a 900 MHz headphone by Sony (http://www.sonystyle.com). The complete kit includes a transmitter, headphones, connecting cord, plug adapter, and AC power supply. The transmission distance is 50 m and it is suitable for listening anywhere in the house and outside.

Wireless computer products are another example of consumer devices. The range of products includes keypads, presentation devices, mice, printers, docking devices for games, and routers. For example, the DSM-320 wireless media player, by D-Link (http://www.dlink.com), can be used to play music, watch videos, or display photos. This device receives data wirelessly from a PC using 802.11g or Ethernet. It supports all popular digital media formats and remote control televisions.

Wireless sports activity monitors are finding extensive applications in the consumer market. An example is illustrated in Figure 5.12. The HR233 is a

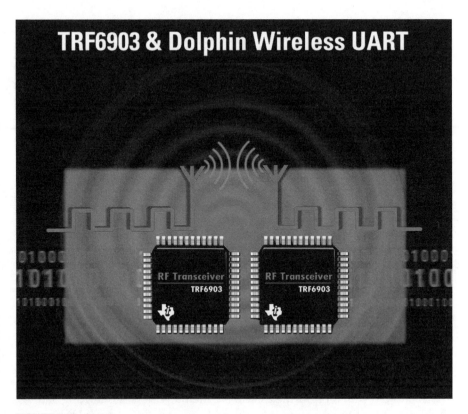

FIGURE 5.12
A wireless pulse rate monitor. (Courtesy of Oregon Scientific: http://www2.oregonscientif-ic.com/catalog/3_9_492.asp)

heart rate monitor produced by Oregon Scientific (http://www2.oregonsci-entific.com) along with many other wireless devices offered by the company. The HR318 monitors the heart rate continuously by a wireless chest strap that transmits information to an accompanying wireless wristwatch. It measures heart rate from 30 to 240 beats per minute and sets upper and lower heart rate limits automatically. It also displays the calories consumed and the fat burning percentage.

Home and office applications provide numerous commercial opportunities for the manufacturing and marketing such devices. Many companies, large and small, offer wireless home security systems. A typical example is the Motorola Home Monitoring and Control System (http://broadband.motor-ola.com/consumers/home_monitoring.asp), which comes as a kit that can be installed by the user. The system consists of a base station/gateway, cameras, door and window sensors, temperature sensor, and water sensor. It can accommodate up to six cameras and up to eight sensors to monitor access points, room temperature extremes, and so on. The system can be set remotely and can send still images and text alerts to a cell phone or e-mail.

A PC, digital subscriber line (DSL), or broadband service is required for settings and notifications via cell phone or e-mail.

The Shell HomeGenie (http://www.shell.com) is another example of a wireless home security system. This system can be used for control purposes in addition to monitoring. It can accommodate up to 32 power switches, 4 wireless and 3 wired cameras, and 2 thermostats. As the company develops additional devices, the system will be able to accommodate a total of 256 devices. The wireless camera offers good quality transmission at a distance of up to 20 m. The distance of transmission depends on the number of obstacles, such as metal framing, thick walls, and other wireless devices operating on 2.4 GHz. The manufacturer suggests that for best results in signal transmission and reception, varying wireless sensor positions and antenna adjustments may be necessary. This system operates on a PC (Pentium II, Windows 98, 128 MB RAM, Internet Explorer 5.5) or Mac (Power Mac G3, OS X 10.2, 128 MB RAM, Safari 1.0).

5.7.2 Other Wireless Applications

Wireless sensor and instrument systems are gradually replacing wired systems in all types of applications. In addition, new applications are being found daily where the use of wires had been impossible or impractical. In this section, further application examples that may not be regarded as unusual in previous sections are provided.

Structural health monitoring (SHM) is a method used for detecting damage and estimating the remaining life of buildings and structures. SHM focuses on large structures such as buildings, bridges, aircraft, ships, spacecrafts, and so on. It offers a paradigm shift from schedule-driven maintenance to condition-based maintenance, hence eliminating manual inspections in lieu of condition-based maintenance for more efficient design practices and more accurate component repairs and replacements. Today, wireless sensor network-based SHM systems are used extensively because of the advantages they offer, including low deployment and maintenance costs, large physical coverage, distributed monitoring, high resolution, and better accuracy.

In the implementation of SHM, adverse changes in a structure are monitored and interpreted in order to reduce life-cycle costs, improve reliability, and implement maintenance programs. Most systems use commercially available sensors and components, such as MEMS sensors, wireless transceivers, piezoelectric actuators and sensors, microprocessors, data loggers, and function generators. In some SHM implementations, sensors are built into the structure or are permanently attached to continuously monitor for structural damage, cracks, deformations, and delaminations, while in other implementations the sensors are attached to an existing structure. The sensor network on the structure provides crucial information regarding the condition, damage state, and service environment of the structure. Diagnostic information from sensor data is used to determine the health of the structure and facilitates an informed decision-making process with respect to inspection and repair.

As an example, a low-cost wireless sensing unit has been designed and fabricated for deployment as the building block of a wireless SHM system by Lynch et al., 2004 . In their study, Lynch et al. deal many issues of wireless SHM systems. The authors state that finite operational lives of portable power supplies necessitate optimization of the wireless sensing unit for overall energy efficiency. This creates a conflict with the need for wireless radios that have far-reaching communication ranges that require significant amounts of power. As a result, a penalty is imposed for transmitting raw time-history records using scarce system resources. The authors also suggest alternatives by using a computational core that can accommodate local processing of data designed and implemented in the wireless sensing unit. The role of the computational core is to perform interrogation tasks on collected raw time-history data and transmit the results of the analysis rather than the time-history data. To illustrate the ability of the computational core to execute such embedded analyses, a two-tiered time-series damage detection algorithm has been implemented. Using a lumped-mass laboratory structure, local execution of the embedded damage detection method has been shown to save energy by avoiding utilization of the wireless channel to transmit raw time-history data.

In another example, Smart Energy aims, on a societal scale, to improve the efficiency of the energy-provision chain. The energy chain consists of three components: generation, distribution, and consumption. Many utility companies are attempting to make the energy supply chain part of an integrated network of monitoring, information processing, controlling, and actuating devices. Some of these processes are wireless.

The goal of wireless sensor and instrument networks in smart energy applications is to improve management and utilization in the generation, distribution, and consumption of electrical energy. At the moment, many organizations are installing wireless systems to ensure immediate contact between field devices. These organizations are able to perform a one-stop energy audit and to instantly produce an analysis of energy consumption for their customers.

In a white paper, the Center for Information Technology Research in the Interest of Society (CITRIS) at the University of California, Berkeley discusses their smart energy projects (Rabaey et al., 2005). The project's goal is to build a societal-scale smart energy system. They predict that instrumenting buildings with a network of tiny and inexpensive wireless electronic sensors could save the State of California as much as $7 billion to $8 billion a year in energy costs. To demonstrate the CITRIS approach, researchers at Berkeley developed an array of wireless sensor nodes and the appropriate software to operate them. They installed matchbox-size "Smart Dust motes" in a building housing the Department of Electrical Engineering and Computer Sciences.

In another example, Smart kindergarten is an interesting application of wireless sensors and instruments (Chen et al., 2002). This is a sensor-based wireless network that targets developmental problem-solving environments for early childhood education. Spatially dense but unobtrusive sensors continuously capture interactions among students, teachers, and common classroom objects. The sensors deliver observations wirelessly to a wired infrastructure for analysis and storage. Two crucial building blocks of this environment are Sylph, a sensor middleware infrastructure, and iBadge, a lightweight sensor-instrumented badge worn by students and teachers. The system can be extended to wirelessly networked, sensor-enhanced toys and other classroom objects with backend middleware services and database techniques. The researchers explore the challenges arising from an ad hoc distributed structure, unreliable sensing, large scale or density, and the novel sensor data types that are characteristic of such deeply instrumented environments.

Parking meters are becoming a common application of wireless systems. As an example, Digital Payment Technologies (http://www.digitalpay-tech.com), together with Linux (http://linuxdevices.com), offer wireless parking components and networks. In their system, the built-in embedded computer provides network connections for real-time payment processing of parking fees, including online and offline credit card transactions. Landline Ethernet or wireless modem Internet connectivity allows the station to be managed remotely. Inside the meter there is an embedded single-board computer (i.e., Intrinsyc CerfBoard 250). The CerfBoard 250 is based on a 400 MHz Intel PXA250 XScale processor, and includes 32 MB flash and 64 MB SDRAM memory; interfaces for 10/100 Ethernet, USB, RS-232 serial, and 45 digital lines; a battery-backed real-time clock; and a CompactFlash expansion socket. Besides handling the networking and payment processing, the operating system interfaces with sensors that monitor activities and states such as paper supply levels. Vibration sensors respond to attempts to move or damage the payment station, while door sensors trigger an alarm in response to theft or damage to the meter.

Vehicle driving over the Internet is the final example provided in this book. In one study, a system was implemented that allowed the user to drive a vehicle over the Internet (Al-Rousan et al., 2003). This system allows a car to be driven by controlling three main functions: the ignition system, speed, and steering. The driver receives real-time video images from a camera mounted on the car. Other critical information such as the car's speed, lights, and gas level are displayed on a Web page. All control and communication between the car and the server takes place through wireless transmitters and receivers that provide total freedom for the movement of the car.

Incidentally, at this point it is worth mentioning that the FCC has recently allocated 5.9 GHz for an 802.11a-style dedicated short-range communication (DSRC) network for cars. It is expected to be used initially for toll collection and measuring traffic congestion. However, there is considerable research interest in applying this frequency band to many other issues, such as control of driverless and intelligent cars, and interaction between the driver and car while driving (Siewiorek et al., 2002).

5.8 Conclusion

In this chapter, all the information provided in the previous chapters has been put to use. It has been shown that wireless sensor and instrumentation systems attract much attention in both R&D and commercial activities. Many examples of wireless systems have been provided, ranging from consumer products to space exploration.

Bibliography

Akinaga, H., High-sensitive magnetic-field-sensing materials composed of metal/ semiconductor hybrid nanostructures, *International Conference on MEMS, NANO, and Smart Systems*, pp. 134–139, 2003.

Al-Rousan, M., Jalajel, F., Arain, F., and Auf A., Web-based wireless vehicle control system, *10th IEEE International Conference on Electronics, Circuits, and Systems*, 1, pp. 388–391, 2003.

American National Standards Institute, http://www.ansi.org.

Asada, G., Dong, M., Lin, T., Newberg, F., Pottie, G., Marcy, H., and Kaiser, W., Wireless integrated network sensors: low-power systems on a chip. *Proceedings of the 24th IEEE European Solid-State Circuits Conference*, pp. 9–12, 1998.

Basios, C., Kostarakis, P., and Pallis, E., Performance evaluation of different wireless network topologies using the IEEE 802.11 MAC protocol, *WSEAS Transactions on Circuits and Systems*, 2, pp. 90–98, 2003.

Beutel, J., Kasten, O., Mattern, F., Roemer, K., Siegemund, F., and Thiele, L., Prototyping wireless sensor network applications with BTnodes, *Proceedings 1st European Workshop on Sensor Networks*, pp. 323–338, 2004.

Bing, B., *Wireless Local Area Networks: The New Wireless Revolution*, New York, Wiley-Interscience, 2002.

Bishop, R.H. (ed.), *The Mechatronics Handbook*, Boca Raton, FL, CRC Press, 2002.

Boukerche, A., Performance evaluation of routing protocols for ad hoc wireless networks, *Mobile Networks and Applications*, 9, pp. 333–342, 2004.

Brooks, T., Wireless technology for industrial sensor and control networks, *SIcon/01, Sensors for Industry Conference*, pp. 73–77, 2001.

Buttazzo, G.C., Scalable applications for energy-aware processors, *Proceedings of the 2nd International Conference on Embedded Software (EMSOFT 2002)*, 2491, pp. 153–165, 2003.

Caldari, M., Conti, M., Crippa, P., Orcioni, S., Solazzi, M., and Turchetti, C., Dynamic power management in an AMBA-based battery-powered system, *9th IEEE International Conference on Electronics, Circuits, and Systems*, 2, pp. 525–528, 2002.

Chandrakasan, A., R. Min, *et al*, 2002, "Power Aware Wireless Microsensor Systems".

Chen, A., Muntz, R.R., Yuen, S., Locher, I., Park, S.I., and Srivastava, M.B., A support infrastructure for the smart kindergarten, *IEEE Pervasive Computing*, 1(2), pp. 49–57, 2002.

Chen, W.K., *The Electrical Engineering Handbook*, Burlington, MA, Elsevier, 2005.

Cheng, X., Huang, X., and Du, D., *Ad Hoc Wireless Networking*, London, Kluwer Academic, 2004.

da Silva, J.S., da Silva, J.J., Wanzeller, M.G., and da Rocha Neto, J.S., Monitoring of temperature using smart sensors based on CAN architecture," *15th International Conference on Electronics, Communication and Computers*, pp. 218–222, 2005.

Dickert, F.L., Lieberzeit, P.A., Gazda-Miarecka, S., Halikias, K., and Bindeus, R., Imprinting with chemical sensors: challenges in molecular recognition and universal application, *Symposium G, Molecularly Imprinted Materials*, pp. 71–77, 2003.

di Natale, C., D'Amico, A., and Dario, P., *Sensors and Microsystems*, Hackensack, NJ, World Scientific, 2002.

Dishman, E., Inventing wellness systems for aging in place, *IEEE Computer*, 37(5), pp. 34–41, 2004.

Dorf, D. (ed.), *Engineering Handbook*, 2nd ed., Boca Raton, FL, CRC Press, 2004.

Dowla, F., *Handbook of RF and Wireless Technologies*, Boston, Newnes, 2004.

Duarte, D., Vijaykrishnan, N., Irwin, M.J., and Tsai, Y.-F., Impact of technology scaling and packaging on dynamic voltage scaling techniques, *Proceedings of the 15th Annual IEEE International ASIC/SOC Conference*, pp. 244–248, 2002.

Dunbar, M., Plug-and-play sensor wireless networks, *IEEE Instrumentation and Measurement Magazine*, March, pp. 19–23, 2001.

Dyer, S.A. (ed.), *Survey of Instrumentation and Measurement*, New York, Wiley, 2001.

Edoardo, B., and Bridges, K., The application of remote sensor technology to assist the recovery of rare and endangered species, *Journal of High Performance Computing Applications*, 16, pp. 315–324, 2002.

Edoardo, B., and Sasaki, G., Wireless sensor placement for reliable and efficient data collection, *Hawaii International Conference on System Sciences*, 2003

Engin, M., Demirel, A., Engin, E.Z., and Fedakar, M., Recent developments and trends in biomedical sensors, *Measurement*, 37, pp. 173–188, 2005.

Eren, H., *Electronic Portable Instruments: Design and Applications*, Boca Raton, FL, CRC Press, 2004.

Fang, Z., Bensaou, B., and Yuan, J., Collision-free MAC scheduling algorithms for wireless ad hoc networks, *IEEE Global Telecommunications Conference*, 5, pp. 2770–2774, 2004.

Fiedler, P., Bradac, Z., Bradac, F., Prokop, M., and Wagner, M., Further evolution of fieldbuses, *WSEAS Transactions on Computers*, 2, pp. 477–480, 2003.

Fraden, J., *Handbook of Modern Sensors: Physics, Design and Applications*, 3rd ed., New York, Springer, 2004.

Frenzel, L.E., Wireless industrial networks: untethered monitoring and control, *Electronic Design*, 52(21), pp. 46–48, 50, 52, 54, 56, 58, 2004.

Gambini, F., The portability of electronic apparatus requires DC/DC converters, *Electronica Oggi*, 321, pp. 104–109, 2003.

Gang, Q., What is the limit of energy saving by dynamic voltage scaling? *IEEE/ACM International Conference on Computer Aided Design*, pp. 560-3, 2001.

Girson, A., Handheld devices, wireless communications, and smart sensors: what it all means for field service, *Sensors*, 19(1), pp. 16–20, 2002.

Graham, I., *Communications*, London, Hodder Wayland, 2001.

Gustafsson, R., Mohammed, A., and Claesson, I., A combined channel estimation algorithm for coherent detection in mobile communication systems, *Wireless 2002, Fourteenth International Conference on Wireless Communication*, pp. 295–299, 2002.

Halgamuge, M.N., Guru, S.M., and Jennings, A., Energy efficient cluster formation in wireless sensor networks, *10th International Conference on Telecommunications*, 2, pp. 1571–1576, 2003.

Haus, C., and Albrecht, H., Neutral data access to fieldbuses in process industries, *IEEE International Workshop on Factory Communication Systems*, pp. 369–372, 2004.

Haykin, S., *Communication Systems*, New York, Wiley, 2001.

Haykin, S., and Moher, M., *Modern Wireless Communications*, Upper Saddle River, NJ, Pearson Prentice Hall, 2005.

Held, G., *Data Over Wireless Networks: Bluetooth, WAP, and Wireless LANS*, New York, McGraw-Hill, 2001.

Hill, J., Szewcyk, R., Woo, A., Culler, D., Hollar, S., and Pister, K., System architecture directions for network sensors, *Architectural Support for Programming Languages and Operating Systems (ASPLOS 2000)*, pp. 93–104, 2000.

Holtzman, J.M., and Zorzi, M., *Advances in Wireless Communications*, New York, Kluwer Academic, 2002.

Horak, R., *Communications Systems and Networks*, Foster City, CA, M&T Books, 2000.

Illman, P.E., *Communications*, New York, McGraw-Hill, 2001.

Ilyas, M., *The Handbook of Ad Hoc Wireless Networks*, Boca Raton, FL, CRC Press, 2003.

International Electrotechnical Commission, http://www.iec.org.

International Organization for Standardization, http://www.iso.org.

Jaewook, R., Yun, S., Kim, B., Park, J.-O., Design and fabrication of a large-deformed smart sensorized polymer actuator, *IEEE/RSJ International Conference on Intelligent Robots and Systems (IROS)*, pp. 908–912, 2004.

Kachirski, O., and Guha, R., Intrusion detection using mobile agents in wireless ad hoc networks, *Proceedings IEEE Workshop on Knowledge Media Networking*, pp. 153–158, 2002.

Katz, D., and Gentili, R., Dynamic power status management of digital signal processors, *Elektronik Industrie*, 33(11), pp. 40–42, 2002.

Kawadia, V., Narayanaswamy, S., Rozovsky, R., Sreenivas, R.S., and Kumar, P.R., Protocols for media access control and power control in wireless networks, *Proceedings of the 40th IEEE Conference on Decision and Control*, 2, pp. 1935–1940, 2001.

Kobak, N.N., Litvin, A.S., and Slesarenko, S.S., The analysis of increase a noise stability by wireless information system, *14th International Crimean Conference on Microwave and Telecommunication Technology*, pp. 263–264, 2004.

Kraver, K.L., Gutthaus, M.R., Strong, T.D., Bird, P.L., Cha, G.S., Hold, W., and Brown, R.B., A mixed-signal sensor interface microinstrument, *Solid State Sensor and Actuator Workshop*, Hilton Head Island, South Carolina, pp. 14–17, 2000.

Kumar, A., and Rahman, F., System for wireless health monitoring, *Proceedings of the ISA/IEEE Sensors for Industry Conference*, pp. 207–210, 2004.

Kurose, J.F., and Ross, W., *Computer Networking: A Top Down Approach Featuring the Internet*, Boston, Addison Wesley, 2005.

Kuruvila, J., Nayak, A., and Stojmenovic, I., Progress-based localized power and cost aware routing algorithms for ad hoc and sensor wireless networks, *Third International Conference on Ad Hoc, Mobile and Wireless Networks*, pp. 294–299, 2004.

Lee, K., Sensor networking and interface standardization, *IEEE-IMTC Conference Proceedings*, pp. 147–152, 2001.

Lee, P.J. (ed.), *Engineering Superconductivity*, New York, Wiley, 2001.

Levanon, N., Implementing Orthogonal Binary Overlay on a Pulse Train Using Frequency Modulation, *IEEE Transactions on Aerospace & Electronic Systems*, 41, pp. 372–382, 2005.

Liptak, B. (ed.), *Instrumentation Engineers Handbook*, 4th ed., Boca Raton, FL, CRC Press, 2002.

Lynch, J.P., Sundararajan, A., Law, K.H., Kiremidjian, A.S., and Carryer, E., Embedding damage detection algorithms in a wireless sensing unit for operational power efficiency, *Smart Materials and Structures*, 13, pp. 800–810, 2004.

Mainwaring, A., Polastre, J., Szewczyk, R., Culler, D., and Anderson, J., Wireless sensor networks for habitat monitoring, *ACM International Workshop on Wireless Sensor Network and Applications*, 2002.

Mallick, M., *Mobile and Wireless Design Essentials*, Indianapolis, Wiley, 2003.

Mani, B., *et al*, 2001 "Smart Kindergarten: Sensor-Based Wireless for Smart Developmental Problem-Solving Environments," *Mobile Computing and Networking*, pp. 132–138.

Matsunaga, H., and Hirano, K,. Modulation/demodulation scheme for high speed modem using matrix and inverse-matrix, *47th Midwest Symposium on Circuits and Systems*, pp. 169–172, 2004.

McDermott-Wells, P., Bluetooth scatternet models, *IEEE Potentials*, 23(5), pp. 36–39, 2005.

McDermott-Wells, P., What is Bluetooth, *IEEE Potentials*, 23(5), pp. 33–36, 2005.

McLean, C., and Wolfe, D., Intelligent wireless condition-based maintenance, *Sensors*, June, 2002, http://www.sensorsmag.com/articles/0602/14/main.shtml.

Molisch, A.F., *Wideband Wireless Digital Communications*, Upper Saddle River, NJ, Prentice Hall, 2001.

Morais, D., *Fixed Broadband Wireless Communications: Principles and Practical Applications*, Upper Saddle River, NJ, Prentice Hall, 2004.

Muller, N.J., *LANs to WANs: The Complete Management Guide*, Boston, Artech House, 2003.

Muller-Fiedler, R., and Knoblauch, V., Reliability aspects of microsensors and micro-mechatronic actuators for automotive applications, *Microelectronics Reliability*, 43, pp. 1085–1097, 2003.

Nakamoto, T., Sukegawa, K., and Sumitomo, E., Higher order sensing using QCM sensor array and pre-concentrator with variable temperature," *IEEE Sensors Journal*, 5, pp. 68–74, 2005.

National Standards Services Networks, http://www.nssn.org.

Nicoletti, S., Zampolli, S., Elmi, I., Dori, L., and Severi, M., Use of different sensing materials and deposition techniques for thin-film sensors to increase sensitivity and selectivity, *IEEE Sensors Journal*, 3, pp. 454–459, 2003.

Noll, A.M., *Principles of Modern Communications Technology*, Boston, Artech House, 2001.

Ohrtman, F., and Roeder, K., *WiFi Handbook: Building 802.11b Wireless Networks*, New York, McGraw-Hill, 2003.

Okamoto, G.T., *Smart Antenna Systems and Wireless LANs*, New York, Kluwer Academic, 2002.

Olexa, R., *Implementing 802.11, 802.16 and 802.20 Wireless Networks: Planning, Troubleshooting and Operations*, Boston, Newnes, 2004.

Ozkul, T., Teaching fieldbus standards to computer engineering students, *IEEE Transactions in Education*, 48, pp. 11–15, 2005.

Packebush, P., Selecting a multifunction data acquisition board for a computer-based measurement system, *Proceedings of the 20th IEEE Instrumentation Technology Conference*, pp. 935–939, 2003.

Panko, R.R., *Business Data Networks and Telecommunications*, Upper Saddle River, NJ, Pearson/Prentice Hall, 2005.

Pardoe, T., and Snyder, G.F., *Network Security*, Albany, NY, Thomson/Delmar, 2005.

Pattan, B., *Robust Modulation Methods and Smart Antennas in Wireless Communications*, Upper Saddle River, NJ, Prentice Hall, 2000.

Petrioli, C., Basgni, S., and Chlamtac,I., Configuring BlueStar: multihop scatternet formation for Bluetooth networks, *IEEE Transactions on Computers*, 52, pp. 779–790, 2003.

Pfleeger, C.P., and Pfleeger, S.L., *Security in Computing*, 3rd ed., Upper Saddle River, NJ, Prentice Hall, 2003.

Poor, R., Wireless mesh networks, 2003, http://www.sensormag.com/article/0203/38/main.shtml.

Pozar, D.M., *Microwave and RF of Wireless Systems*, New York, Wiley, 2001.

Rabaey, J., Arens, E., Federspiel, C., Gadgil, A., Messerschmitt, D., Nazaroff, W., Pister, K., Oren, S., and Varaiya, P., Smart energy distribution and consumption information technology as an enabling force, *CITRIS White Paper*, 2005, http://www.citris.berkeley.edu/SmartEnergy.

Raghavendra, C.S., Sivalingam, K.M., and Znati, T., *Wireless Sensor Networks*, London, Kluwer Academic, 2004.

Raja, P.C., and Suganthi, K, VLSI approach to wireless security mechanisms, *IEEE International Conference on Personal Wireless Communications*, pp. 429–433, 2005.

Ramjee, R., La Porta, T., Salgarelli, L., Thuel, S., Varadhan K., and Li, L. IP-based access network infrastructure for next-generation wireless data networks, *IEEE Personal Communications*, 7(4), pp. 34–41, 2000.

Rappaport, T.S., *Wireless Communications: Principles and Practice*, Upper Saddle River, NJ, Prentice Hall, 2002.

Reynolds, J., *Going Wi-Fi: A Practical Guide to Planning and Building an 802.11 Net*, Gilroy, CA, CMP Books, 2003.

Rong, P., and Pedram, M., Battery-aware power management based on Markovian decision processes, *IEEE/ACM International Conference on Computer Aided Design*, pp. 707–713, 2002.

Ross, P.E., Managing care through the air, *IEEE Spectrum*, December, pp. 14–19, 2004.

Ryan, M.J., and Frater, M.R., *Communications and Information Systems*, Canberra, New South Wales, Australia, Argos Press, 2002.

Santamaria, A., and Lopez-Hernandez, F., *Wireless LAN Standards and Applications*, Boston, Artech House, 2001.

Schwiebert, L., Gupta, S.K.S., Auner, P.S.G., Abrams, G., Lezzi, R., and McAllister, P., Biomedical smart sensor for the visually impaired, *IEEE Sensors*, 1, pp. 693–698, 2002.

Shafi, M., Ogose, S., and Hattori, T., *Wireless Communication in the 21st Century*, New York, Wiley-Interscience, 2002.

Siegemund, F., and Krauer, T., Integrating handhelds into environments of cooperating smart everyday objects, *Proceedings 2nd European Symposium on Ambient Intelligence (EUSAI 2004)*, pp. 160–171, 2004.

Siewiorek, D., Smailagic, A., and Hornyak, M., Multimodal contextual car-driver interface, *Proceedings Fourth IEEE International Conference on Multimodal Interfaces*, pp. 367–373, 2002.

Sinha, A., Wang, A., and Chandrakasan, A., Energy scalable system design, *IEEE Transactions on VLSI Systems*, pp. 135–145, 2002.

Sklar, B., *Digital Communications: Fundamentals and Applications*, Englewood Cliffs, NJ, Prentice Hall, 2001.

Smith, D.R., *Digital Transmission Systems*, Boston, Kluwer Academic, 2004.

Smith, P., *Mobile and Wireless Communications: Key Technologies and Future Applications*, Stevenage, UK, Institution of Electrical Engineers, 2004.

Snoonian, D., Smart buildings, *IEEE Spectrum*, August, pp. 18–23, 2003.

Sriram S., Reddy, T.B., Manoj, B.S., and Murthy, C.S.R., The influence of QoS routing on the achievable capacity in TDMA-based ad hoc wireless networks," *IEEE Global Telecommunications Conference*, 5, pp. 2909–2913, 2004.

Srivastava, M., Muntz, R., and Potkonjak, M., Smart kindergarten: sensor-based wireless networks for smart developmental problem-solving environments, *Proceedings of the 7th International Conference on Mobile Computing and Networking*, New York, ACM Press, pp. 132–138, 2001.

Stallings, W., *Data and Computer Communications*, 7th ed., Upper Saddle River, NJ, Pearson, 2004.

Sun, Y., *Wireless Communication Circuits and Systems*, London, Institution of Electrical Engineers, 2004.

Sydenham, P.H., and Thorn, R. (eds.), *Handbook of Measuring System Design*, New York, Wiley, 2005.

Takahashi, S., Hamamura, M., and Tachikawa, S.A., Demodulation complexity reduction method using M-algorithm for high compaction multi-carrier modulation systems, *1st International Symposium on Wireless Systems 2004*, pp. 418–422, 2004.

Tan, G., and Guttang, J., Locally coordinated scatternet scheduling algorithm, *27th IEEE Conference on Local Computer Networks*, 2002.

Tanenbaum, A.S., *Computer Networks*, Upper Saddle River, NJ, Prentice Hall, 2003.

Temple, R., and Regnault, J., *Internet and Wireless Security*, London, Institution of Electrical Engineers, 2002.

Thomas, L., Kalinowski, J., Cook, C., and Verduzco, L., Fieldbus instruments support a larger vision: intelligence moves plant from preventive maintenance to predictive maintenance, *Intech*, 51(7), pp. 22–26, 2004.

Tomasi, W., *Electronic Communications Systems: Fundamentals Through Advanced*, Englewood Cliffs, NJ, Prentice Hall, 2001.

Tomsho, G., Tittel, E., and Johnson, D., *Guide to Networking Essentials*, 3rd ed., Thomson Learning, Toronto, Ontario, Canada, 2003.

Vaidyanathan, R., Kant, L., McAuley, A., and Bereschinsky, M., Performance modeling and simulation of dynamic and rapid auto-configuration protocols for ad hoc wireless networks, *Proceedings 36th Annual Simulation Symposium*, pp. 57–64, 2003.

Villarruel, J.G., Gómez, D.S., González Pérez, L.F., and García, A.G., RF analysis applied to digital wireless receivers design, *15th International Conference on Electronics, Communication and Computers*, pp. 36–41, 2005.

Voglewede, P.E., and Hessel, C., Efficient error free data transmission in a line of sight bandlimited channel, *MILCOM 2000, 21st Century Military Communications. Architectures and Technologies for Information Superiority*, pp. 266–270, 2000.

Wang, K., Chiasserini, C.-F., Proakis, J.G., and Rao, R.R., Distributed fair scheduling and power control in wireless ad hoc networks, *IEEE Global Telecommunications Conference*, 6, pp. 3556–3562, 2004.

Wang, S.Q., MacDonald, P., Kruger, M., and Welch, M., Wafer level wireless temperatures sensing and its applications in RF plasma etch in VLSI processing, *Asia-Pacific Radio Science Conference Proceedings*, pp. 490–493, 2004.

Webster, J.G. (ed.), *Electrical Measurement, Signal Processing, and Display*, New York, Wiley, 2004.

Webster, J.G. (ed.), *Mechanical Variables Measurement: Solid, Fluid, and Thermal*, New York, IEEE Press, 2000.

Weisman, C.J., *The Essential Guide to RF and Wireless*, Upper Saddle River, NJ, Prentice Hall, 2002.

Wismeijer, D., Customizing fieldbus function blocks, *Control Engineering*, 51, pp. IP1–4, 2004.

World Standards Services Networks, http://www.wssn.net.

Worms, P., Security reaches the end of the line, *Network Computing*, 13(7), pp. 10–11, 2004.

Wu, H., Huang, C., Jia, X., and Bai, B., QoS routing of multiple parallel paths in TDMA/CDMA ad hoc wireless networks, *Proceedings of SPIE*, 5284, pp. 205–213, 2004.

Xiaoning, Y., et al., Investigation of PCB layout parasitic in EMI filtering of I/O lines," *IEEE EMC International Symposium*, pp. 501–504, 2001.

Xinxia, C., and Dafu, C., The development of biosensors and biochips in IECAS, *Network and Parallel Computing*, IFIP International Conference, pp. 522–525, 2004.

Xu, N., A survey of sensor network applications, 2003, http://enl.usc.edu/ningxu/papers/survey.pdf.

Ying, H., and Chong, K.P., Sensor scheduling for target tracking in sensor networks, *43rd IEEE Conference on Decision and Control*, pp. 743–748, 2004.

Yuan, L., and Gang, Q., Design space exploration for energy-efficient secure sensor network, *Proceedings IEEE International Conference on Application-Specific Systems, Architectures, and Processors*, pp. 88–97, 2002.

Zaruba, G.V., Basagni, S., and Chlamtac, I., BlueTrees-scatternet formation to enable Bluetooth-based personal area networks, *Proceedings of the IEEE International Conference on Communications*, 2001.

Zhang, Y., and Chakrabarty, K., Energy-aware adaptive checkpointing in embedded real-time systems, *Design, Automation and Testing in Europe Conference and Exhibition*, pp. 918–923, 2003.

Ziemer, R., and Peterson, R., *Introduction to Digital Communications*, Englewood Cliffs, NJ, Prentice Hall, 2001.

Ziemer, R.E., and Tranter, W.H., *Principles of Communications: Systems, Modulation, and Noise*, New York, Wiley, 2002.

Index